FAMILY SURVIVAL MEDICINE HANDBOOK

Essential First Aid & Emergency Practices
Every Family Should Know When
Help Is Not on the Way

The Wilderness Medical Protection Plan for the Non-Medically Trained

By Survival Knowledge Is Power
www.SkipPressOfficial.com

Family Survival Medicine Handbook
Essential First Aid & Emergency Practices Every Family Should Know When Help Is Not on the Way

Published by Survival Knowledge Is Power (SKIP) Press.

Copyright ©2022 Survival Knowledge Is Power (SKIP) Press. All rights reserved.

No part of this book may be reproduced in any form or by any mechanical means, including information storage and retrieval systems without permission in writing from the publisher/author, except by a reviewer who may quote passages in a review.

All images, logos, quotes, and trademarks included in this book are subject to use according to trademark and copyright laws of the United States.

SURVIVAL KNOWLEDGE IS POWER (SKIP) PRESS, Author

FAMILY SURVIVAL MEDICINE HANDBOOK ESSENTIAL FIRST AID AND EMERGENCY PRACTICES EVERY FAMILY SHOULD KNOW WHEN HELP IS NOT ON THE WAY

ISBN: 979-8-88538-002-7

All rights reserved by Survival Knowledge Is Power (SKIP) Press.
www.SkipPressOfficial.com

"There's no harm in hoping for the best as long as you're prepared for the worst."

— **Stephen King**

A Special Gift to Our Readers

Included with your purchase of the book is our SKIP Press Coupon Code Booklet + ***bonus*** Survival Medicine Trivia Game. Our coupon code booklet will give you discounts just for being part of our community! The game, based upon the book, will help you learn important life-saving measures, foundational medical care for when you cannot access help, mental preparation, and essential survival knowledge to survive any disaster.

Simply click or visit the link to save and play!
www.SkipPressOfficial.com

Contents

Introduction   vii

1. Survival Medicine: All You Need to Know   1
2. Preparing Towards Emergencies   13
3. Taking Control of the Situation: Initial Assessment   25
4. Potential Body Injuries and Treatment   75
5. Useful Techniques for Health Emergencies   105
6. Handling Minor Health Emergencies   145
7. Handling Severe Health Emergencies   165
8. Handling Health-Related Emergencies   183
9. Environmental Emergencies   207
10. Where to Find Shelter in Extreme Weather   221
11. Medications: The Alternatives   267

Conclusion   285
References   287
Index   305

INTRODUCTION

Life is easy if the world is in order: we have round-the-clock electricity, running water, internet, and telephone service. You can order food, clothes, medicine, and other supplies within a few seconds via online services. Hiring a vehicle takes only a minute. Emergency medical services can reach you quickly when you need them. The world as you know it, at present, literally functions at the click of a button or the tap of a cellphone.

But what happens if you suddenly find yourself without these things? What if you take your family out on a wilderness hike where there is no cellphone signal and there are no medical facilities nearby, then someone injures themselves? What if, during another pandemic, health services become so strained and overcrowded that you can't get professional medical help when you or your family need it?

Scenarios such as these might seem unlikely, but they are not so difficult to imagine. We don't know when something outside of our control, such as terrorist activities, civil unrest, natural calamities, or disease outbreaks—as seen in recent times of COVID-19—could cause the services we normally rely upon to collapse. We don't know when we might find ourselves off-grid and needing to rely on our own knowledge and skills. In such times, if you or one of your closest family members or friends need immediate medical assistance, then the ability to survive hinges on your preparedness.

In situations like these, wouldn't you want to be as prepared as possible? Imagine being able to confidently help burn victims or manage a soft tissue injury, altitude sickness, or other dire emergencies. Imagine being able to use natural plant extracts when you don't have pain medications or antiseptics. Imagine being able to locate and distinguish healing plants in the woods.

Are you not as prepared as you would like to be when it comes to emergencies where life as we know it stops? Or are you perhaps just starting on your survival prep journey and you do not know where to begin? Or have you been living this lifestyle for years and are seeking to increase your book arsenal? If you answered yes to any of these questions, this book is for you.

We understand that many of you may know how to provide first aid to a patient in certain scenarios. We have all been taught to:

1. Analyze the situation
2. Attend to and stabilize the patient's condition
3. Transport the patient to a nearby health facility

But in this book, we are talking beyond these principles. We want to explain what to do if the third step is not possible, i.e., if you are unable to transport yourself or the patient and will have to be the medical expert yourself. You surely do not want to freeze at that point due to panic and unpreparedness. The rescue helicopter may not be there for you in a few hours, a few days, or even a week. And in such a case, we know that you want to feel confident in having the basic medical knowledge to provide assistance to your close and dear ones.

We have all recently witnessed a pandemic. Because of personal experiences and those of close friends and family, we have learned that we need to be more prepared next time. This series of survival books will help us all be better prepared for the days to come, whether they bring another wave of COVID-19, natural calamities, civil unrest, or outdoor accidents. It could end up making the difference between life and death.

This particular book in the series will address the medical skills we all should have in case of an emergency. We can prepare for common ailments and nasty surprises that could occur in our homes or on remote outdoor trips. This book is for families to have in their homes to use as a starter guide in case of unpredictable events. We have tried to answer your concerns about:

- Measures to medically treat yourself or others when help is not on the way
- Items that you should get now to have every day at home and to bring when you go camping, hiking, or exploring
- The step-by-step guide to preparedness, both physical and mental

- The medical training you need to resolve common ailments
- Some natural healing remedies that are readily available in nature

Let's be prepared for worse days in a way everybody can understand. With each other and our expertise, we can help stop a common ailment from progressing to a life-threatening one during a disaster. We can alleviate physical discomfort and we can potentially save lives! That said, some ailments or injuries can reach a point of severity where the next step is to determine how to transport the patient to a nearby physician or get a physician to you. In those situations, that is the best use of your energy. We will do our best to explain how to prevent such emergencies and what to do in the meantime, but, unfortunately, some situations get to that point.

This book is not full of medical jargon. Instead, it is written for everyday people with a non-medical background so that you can understand and actually learn without having to memorize a whole bunch of steps. This will also assist you in being resourceful. That said, some information is repeated in an effort to help you learn and make this book more user friendly.

How to Use This Book

In this book, you can quickly use the following method to find the information you need:

- First, look at the table of contents at the beginning of

the book. The names of the chapters will provide you with an idea of the content within.
- Second, once in the chapter, note that there may be illustrations that can answer your questions very quickly. These can give you an overview of the contents elaborated on in the chapter.
- Third, if those options fail, there is an index listed on the end pages of the book. The index has words with their corresponding page numbers. Go through these for a quick search of the related content.

Chapter 1 deals with the mental preparations you may want to go through before beginning with the actual treatment process. Chapter 2 gives you information on how to make your first-aid kit and other measures for practical preparedness, along with mental readiness. Chapter 3 is about initial assessment and taking control of the situation. Chapters 4-10 are related to various potential emergencies that can be encountered, along with their management. And finally, if you want to know which other materials can be used as an alternative to conventional medicine, you can go through Chapter 11. Have a great time learning and always stay prepared!

Disclaimer

This book is not intended to be an all-in-one, in-depth medical survival guide. However, it will jump-start your medical survival preparedness journey and serve as a great addition to your survival library. Our SKIP team has over 15 years of combined medical experience. This book was written to try to help others before it's too late. However, all the contents of this book are for informational purposes only. Do not substitute this book for professional medical advice or delay treatment because of something you read in this book.

Always consult your doctor or other qualified health personnel for any health issues and be sure to call emergency medical services in emergencies, whenever possible. Also, we highly recommend you further your medical preparedness by taking training courses to physically prepare yourself for if or when you are put in this situation. These training sessions will help you practice some of the tools in this book so they become second nature. Some training and certifications you should consider are wilderness first-aid courses (2 days) or wilderness first responder courses (4-5 days).

1
SURVIVAL MEDICINE: ALL YOU NEED TO KNOW

A group of us were in our early twenties when we first received an invitation to a survival medicine seminar. The seminar started with questions on what survival medicine was. The first thing that came to mind was an emergency in the household where we had to keep the patient alive until the ambulance arrived. Also, there were mentions of an unexpected scenario in the middle of the forest while camping. In both of the cases, help was only a few minutes away, and first aid was to be given until professional help was available. It became apparent that we were not the only ones who thought this way. Most of the other attendees answered almost the same. Scenarios of an earthquake, fire accident, or sudden loss of consciousness were imagined and portrayed. We had confused survival medicine with the initial first aid administered to any sick and dying patient. However, what we learned afterward left us with a new outlook on survival medicine.

Let us explain to you the things we learned back then. We intend to explain the concept of survival medicine clearly and also differentiate it from basic first-aid care. I would like to start with a detailed description of survival medicine. Let me begin by telling you that survival medicine is also known as "wilderness medicine" or "outdoor medicine."

What Is Survival Medicine?

Survival medicine is the care provided in an environment where the standard medical facilities or equipment are beyond reach. It may be in a remote wilderness where medical facilities are not accessible. Or it may be in a disease-stricken city where the facilities are understaffed, undersupplied, and overloaded such that they cannot accommodate any more patients. This mainly occurs in instances of civil unrest, terrorist activities, natural disasters, or pandemics like that of COVID-19, which we have recently witnessed.

There are three independent factors in survival medicine:

- An austere or unfriendly and hostile environment with no comfort or luxury whatsoever—e.g., jungle, water, heat, cold, or high altitude. This may either be in a remote setting or a disaster-stricken urban area.
- Inadequate resources with no or suboptimal medical supplies. The resources in the setting are either unavailable or insufficient. Natural resources may be the only thing left for our use.
- Substantial delay in access to medical care and resources. Health facilities may be a far-fetched resource in wilderness medicine. Help is significantly delayed and you are the only human resource available at the moment!

This book will seek to address these particular scenarios and their management.

Survival medicine focuses on providing care in hard-to-reach environments. The people who respond to these environments are known as wilderness responders, remote care specialists, or wilderness Emergency Medical Technicians (EMTs). While this book seeks to give you the basics when it comes to this type of care, it also helps you become more mentally comfortable about how you handle these situations, as one can never be overly prepared when it comes to potential life-or-death situations. In addition to reading this book, we recommend you take wilderness training courses to further enhance your readiness. These courses are special because they offer more of a hands-on experience versus just reading a book. Thus, they further enhance your mental preparedness, which, as we will learn later on, is just as important as physical preparation in managing wilderness emergencies.

For various people, the phrase "wilderness" might signify different things. Traditionally, wilderness refers to: (a) a tract or territory that has not been farmed or inhabited by humans, and (b) an area that has not been disturbed by human activity and is in peace with its natural flora (plant life) and fauna (animal life) (Bowman 2001). However, in modern practice, it not only refers to the geographical representation but also the activities and injuries that come with it.

A fuller definition of wilderness medicine includes the following three concepts (Sholl and Curcio 2004):

- The location where the injury occurred:

    Over the years, the number of people going deep-sea diving, mountain biking, trekking, and snowboarding, as well as enjoying other adventure sports, has been rising. There has been an increased interest and participation in these recreational activities. Previously, some sports like mountaineering and rock climbing were reserved only for the experts. However, they are now also being enjoyed by many amateurs with no professional training. There are risks and injuries associated with these adventures. This further increases the need for knowledge of survival medicine.

Survival medicine provides care with limited supplies. In a remote environment, there may be no electricity or water supply, let alone pharmacy stores. Survival medicine responders must work with the supplies readily available in the daily household and natural environment. These may include clothes and wooden blocks for fracture management or herbs and plant extracts for wounds, and so on.

- Time to definitive care:

Survival medicine can provide long-term care for the injured when traditional help is not available. The length of treatment is determined by the severity of the injury as well as the resources available in the surrounding environment. It should be acknowledged that the scenario has changed from a safe environment to a critical one, and a problem has been encountered. Supplies are now limited, so the available resources (like water, canned foods, and medical supplies) should be made to last longer periods of time. You are the only medical asset in the situation.

- Type of injury:

    Common injuries in wilderness medicine include bruises, muscle pulls, sprains, and lower-extremity fractures. Throughout this book, we will be guiding you to provide basic life-saving measures and regarding how to manage these common injuries in a medical emergency.

Looking back in history (Sholl and Curcio 2004), wilderness medicine was known to address the injuries sustained by the military in war zones. Lessons learned in the treatment and transport of casualties have led to the development of wilderness medicine as we know it today. At present, wilderness medicine is a broad concept involving multiple areas, including the following:

- Trauma
- Cold injuries, including frostbite and hypothermia
- Heat illnesses

Although wilderness medicine is an evolving concept encompassing a number of subjects, in this book, we chose to focus on some of the most common scenarios that can be encountered.

In this book, we will also be dealing with basic clinical examination, life-saving measures, and case-specific scenarios of management in the wilderness. We will provide the basics of medical preparedness for multiple days with or without traditional help.

Survival Medicine Vs. First Aid

The definition of first aid is "the initial care offered for a severe disease or injury." One of the primary purposes of first aid is to save lives. Aside from that, first aid attempts to alleviate pain, prevent additional disease or damage, and improve the chances of healing (Sholl and Curcio 2004).

The general principles of providing first aid to the injured are as follows: recognizing, assessing, and prioritizing the need for first aid; providing care; recognizing limitations in doing so; and looking for additional services when necessary, such as activating emergency medical services or other medical assistance. First aid includes performing a basic head-to-toe examination, assessing vital parameters, treating minor and severe bodily injuries, treating burn injuries, controlling severe allergic reactions, treating bites, and much more.

First-aid response times by ambulance for a 911 call vary. In the US, those who live in the city wait on average 7 minutes, those in more rural areas 14 minutes (Alvarado 2020).

The main aims of providing first aid in an emergency are (Charlton et al. 2019):

- Preserving life
- Preventing deterioration
- Promoting recovery
- Relieving pain
- Protecting the unconscious

First aid and survival medicine share the same aims. Hence, this book intends to teach you the basics of both survival medicine and first-aid administration.

Providing first-aid treatment is a sequence that requires you to (AIP Safety 2021):

- Assess a situation quickly and calmly
- Protect yourself from the dangers of the situation. Another person getting injured is the last thing needed
- Prevent cross-contamination (transmission of germs resulting in harmful effects) between yourself and the injured person
- Comfort the casualty or injured party
- Assess the status of the casualty along with the nature of the illness
- Provide treatment, prioritizing life-threatening situations
- Arrange for appropriate help

This sequence also applies to survival medicine. We will be discussing this sequence in case-specific scenarios in this book.

The concepts of first aid and survival medicine overlap in that they both aim to provide medical care to sick and dying patients. However, in first-aid, help is always on the way. In wilderness medicine, a "first-aid mindset" isn't enough. Help may take longer due to the adversity of the environment. You may have to provide help as long as medical assistance remains inaccessible.

Mentally Preparing for Survival Medicine

Being prepared for any kind of emergency requires forethought about health care. Health care requires technical skills, resources, and preparation. However, what it also requires that many overlook is a stable mind to ensure the skills we do know are properly executed and we do not waste valuable resources, which are likely already limited.

Being prepared for the worst does not mean you are hoping for it. It does not imply a negative outlook on life. Instead, it is all about staying prepared for whatever may come your way. As the saying goes, "Better safe than sorry;" being prepared and equipped can help you assess medical emergencies quickly and accurately and provide appropriate care. You will remain calm and save as much of your energy as possible, which is needed, especially in life-or-death situations. You will be confident instead of panicking and looking for others to help. You will be prepared to help yourself and those in need. And what could be better?

Your mind needs to be calm and prepared if disaster strikes, because your body is already in a stressful environment. You may have injuries with blood dripping out; you may have an animal bite or a fracture. Situations like these can cause your body to go into fight-or-flight mode. That means your body either intends to fight the situation or run from it. In the initial stage of emergency response, hormones, which are chemical substances present in our bodies, stimulate specific cells into action. These hormones can increase your heart rate, raise your blood pressure, and release energy reserves and supply them to your muscles so that you are ready to combat the situation (Conger n.d.).

If you are not mentally prepared, this initial stage can utilize so much of your energy that you become exhausted to the point of not being able to think straight or clearly, becoming lethargic or frozen.

In cases of unfavorable circumstances, you do not want this stress cycle to take you over. You don't want to freeze due to a mental breakdown. You do not want to faint by overwhelming your nervous response. Instead, you want to help yourself. This may mean you need to reverse the damage, be it by descending from a height if you are getting altitude sickness, moving to a safe place if you have strayed too far, or returning to your camp if you hear the weather ahead isn't good, rather than trying your luck. Or it may mean that you want to prevent further damage from occurring by attending to minor injuries. It helps to prepare psychologically for whatever comes your way.

To be psychologically strong in any chaotic situation, you have to (Conger n.d.):

- Address any fears or negative emotions. First, you have to acknowledge that you are feeling scared in the situation. And don't panic! Everybody feels scared when they are out of their comfort zone. Acknowledging this emotion is the first step towards solving the problem causing it.

- Alleviate the fear of the unknown. This is more easily done if you are aware of what to do in the situation. This can be achieved by reading books like this one on survival guides and/or taking first-aid and wilderness courses beforehand.
- Repeat reassurances and positive affirmations to yourself. Constantly repeating things in your mind helps you believe them. So, remind yourself repeatedly that you are brave and you can make it!
- Don't live with blame! Filling your mind with negative emotions can be toxic in the situation. Instead, accept that you made a mistake so that you can move on to figure out how to resolve the problem. Look at the brighter side: you are going to learn from the situation. Move on!
- Stay busy and keep your mind occupied. Use the situation to your advantage. Gather things from around you that may be useful in your management plan. There may be sticks or wooden boards around you, or perhaps some clothes that can help in your treatment if you have a broken limb. Collect those and try cleaning them for use. But remember, you have to save energy, so do not overwork yourself! You may have read about managing similar scenarios or watched some related movies and documentaries before. Try to remember them. They will help you handle the situation.

It is absolutely important to develop a healthy perspective on survival medicine. Disaster preparation should not be undertaken out of sheer fear or any negative emotion. Instead, you need to have a positive outlook on taking care of yourself and those near and dear to you. Become a problem solver who not only survives but thrives in a bad situation!

This book will help you make confident decisions and receive a morale boost when it comes to emergency preparedness.

Takeaway

Survival medicine, also known as "wilderness medicine," deals with the management of a patient in a hard-to-reach environment with limited resources and with no or delayed help. First aid deals with the basic life-saving measures administered to a patient until the ambulance arrives. These two disciplines have a lot in common in terms of patient management. Thus, this book will teach you the basics of both. We intend to help you become more physically and psychologically prepared for some of the most common emergency medical situations.

This is the book that will help you develop the right mentality to perform the necessary activities in an emergency situation when help is not on the way. Survival medicine is your ultimate helpline between life and death out there in the wilderness!

2
PREPARING TOWARDS EMERGENCIES

In 1704, a navy officer was left behind on an uninhabited island by his captain. This officer spent 4 years and 4 months alone before rescue help arrived. When help did finally arrive, he was seen eating spiny lobsters and wild turnips, and he had a shelter made of pepper trees. This is the famous story of Alexander Selkirk. His story was mentioned in many articles and stories published thereafter. *Selkirk: The Real Robinson Crusoe* was an animated movie that premiered in 2012, which described the events in Selkirk's life.

Now, let us look at some other, different scenarios. There have been reports of people being rescued from earthquake rubble after up to 27 days. Also, a lot of us have heard of real-life, at-home stories where a kid saves his or her mom or dad by performing CPR or the tales where families survived ravaging tornados because they all went to that one part of the home that was safe.

These real-life stories occurred in different environments. However, they have certain things in common. All of these were unexpected emergencies and the stories reflect how an individual survives an emergency by their level of preparedness. Had these people lost hope and crumbled in the devastating situation, these stories would never have been told. Your level of physical and mental preparedness and knowledge of how to do things can mean the absolute difference between surviving and not surviving.

Preparedness refers to the systematic mobilization of individuals, equipment, and supplies in a safe atmosphere for effective assistance (Abebe 2009). For example, before the situation occurs, preparation can be done by paying attention to the weather forecast and warning systems for seasonal climate change, attending survival preparedness seminars and trainings, and holding drills and exercises. When preparing in advance, if and when any unexpected emergencies occur, you can ensure your readiness by storing food and water, building temporary shelters, devising management strategies, planning for evacuation and reallocation, and so on.

The aims of emergency preparedness are as follows (Sena and Woldemichael 2006):

- Reduce (or avoid, if possible) the potential losses from hazards. This involves taking measures either to prevent an injury from occurring or to reverse a bodily injury by appropriate treatment.
- Ensure prompt and appropriate assistance when necessary. Time is very crucial when it comes to managing an injury.
- Achieve a rapid recovery. Providing immediate care can be life-saving for an injured person. If appropriate care is provided promptly, rapid recovery can be achieved and potential complications minimized.

Summed up, you should ultimately be able to reduce the chance of immediate death (mortality) and also the chances of any injured person being handicapped for life (morbidity) with a better-prepared approach.

Preparedness includes mental and practical preparedness.

Mental Readiness

We have discussed the basics of mental preparedness in unexpected emergencies in Chapter 1. In this chapter, we will provide you with some more details on how we can use our mindset to the best advantage.

As a recap, our body responds differently in times of stress. Stress is a response to an event that an individual finds physically, mentally, or emotionally challenging. In an emergency situation, stress can occur in the form of physical, mental, or emotional stress, or all three. In addition, the procedure of decision-making itself can be stressful. Many studies have shown that stress has negative effects on decision-making capacity (Porcelli and Delgado 2017; Wemm and Wulfert 2017). It hinders an individual's ability to think clearly and make appropriate choices. This may result in a vicious cycle in which stress affects the decision-making process, which in turn causes more stress in the situation.

This cycle results in the rapid utilization of our body's energy sources. Thus, we end up being mentally fatigued when we need to be working promptly to use the surrounding resources to our advantage. So, how do we avoid this?

Mental readiness is the solution and it need not start when you encounter the situation. It can start today! These are the steps you can take to ensure that you are mentally prepared for the situation:

- Make plans: For example, talk openly about the emergencies you can encounter when you go on outdoor hikes or do adventure sports. Also, discuss where to go and what to do if a hurricane, tornado, earthquake, or other disaster were to threaten your household. Take seminars and share the knowledge among yourselves.
- Learn more about the negative effects of ill-preparedness: As much as it is necessary that you know how to execute plans, it is also necessary to know the consequences if you do not take action in time so that you act promptly. Your decision at the right time can save your life or that of your patient.
- Remember that practice makes perfect: This is also known as emergency conditioning. This is a training technique to make unknown situations seem familiar, tricking your brain into thinking that you have already undergone the situation before. In wilderness medicine preparedness, not everything may be practicable, but you can surely practice basic first aid and positive self-reinforcement.
- Think positively: After you have made plans and practiced skills enough, you are more likely to think positively. Repeat motivating phrases in your mind, such as, "You can do it." This will energize your mind. Thus, you are more likely to handle the situation effectively and get out of trouble (Myers 2017).

Practical Preparedness

After your mind is well set to act in an outdoor emergency, at-home situation, or a disaster-stricken household, you are now ready for action. You are in a situation with no access to supplies for days. Thus, a disaster preparedness kit may be the ultimate help.

Below is the list of essential items that the Federal Emergency Management Agency (FEMA) recommends including in a disaster supply kit. If a disaster strikes or in the case of civil unrest, you may not be able to go outdoors to collect resources like:

- Water: At least one gallon (approx. 3.7 liters) per person per day for several days can be used for drinking and personal hygiene
- Food: Processed or canned food that can be stored for a longer period should be available for at least a three-day supply
- Manual can opener for canned food
- Battery-powered or manual radio to provide updates on the weather with tone alert
- First-aid kit
- Cell phone with chargers and a power bank, if possible
- Flashlight
- Extra batteries
- Whistle (to signal for help)
- Mask (to help filter contaminated air)
- Plastic sheets, duct tape, and rope (to make a temporary shelter)

- Wet wipes, garbage bags, and ties (to maintain personal sanitation)
- Tools to turn off utilities; for example, wrenches or pliers in case of water leakage
- Local maps
- Prescription medications and over-the-counter medications (medicines available without a prescription; e.g., pain killers, antacids, etc.)
- Any special items for the elderly or disabled members of the family (Cunha 2021)

Additionally, we recommend:

- Cash
- Pen and paper for note taking
- Sharp knife for cutting things such as cord or rope, branch removal for fire, or shelter, bandages, and of course, for self-protection

These items need not necessarily be in a "kit" but should be within easy reach. Also, in contrast, when you are hiking outdoors, items such as food and water may not be available in the exact amounts mentioned. This is when you need to search for natural resources in the area so that you can replenish them according to your needs.

These supplies (or others that will accomplish the same goal of providing food, water, shelter, and medical care) are the basic survival tools that will significantly help to ease your difficulties in the event of an emergency wilderness experience.

First-Aid Kit

In the event of a medical emergency, first-aid preparation is also needed.

Here is the list of necessary items you need to have in your first-aid kit when you are preparing for a wilderness experience or to keep in stock at home:

- Adhesive tape – for securing bandages
- Gauze pads (sterile) – for padding bleeding wounds
- Non-adhesive wound pads – for absorption of wound discharge or bleeding and also to minimize injury to the wound site
- Adhesive bandages (all sizes)
- Triangular bandage – a sling, towel, or tourniquet (a band wrapped tightly around hands or legs to prevent the flow of blood for some time).
- Adhesive roll – to apply to blisters or hot spots
- Antacids – for indigestion
- Antidiarrheal agents
- Stool softeners or laxatives – to increase bowel movement and help alleviate constipation
- Cream – for rashes and allergies
- Antiseptic solution (small bottle liquid soap) – for cleaning wounds and hands
- Diphenhydramine (Benadryl) – for allergies and the common cold
- Cough medication

Preparing Towards Emergencies

- Oral rehydration solution (ORS) – for electrolytes in the case of diarrhea
- Pain-killers like ibuprofen or aspirin
- Nasal spray or oral decongestant – for nasal stuffiness caused by colds or allergies
- Oral antibiotics and ointment – to apply to skin wounds for faster healing
- Personal drugs – enough for the duration of the trip or a week at home, and possibly a couple of extras in case of time delays
- Medical items (for example, knee braces)
- PPE, or personal protective equipment – a gown, gloves, mask or face covering, and eye protection to prevent sharing germs when treating patients
- Safety pins (large and small)
- Scissors
- Thermometer
- Tweezers – for removal of small contaminants from a wound, such as splinters or ticks
- Cigarette lighter – to make a fire in the wilderness (to keep warm, boil water for food or drink, to cook, to make smoke to signal for help, and also to sterilize instruments)
- Iodine – to treat water
- List of important people to reach in an emergency
- Paper and writing utensils to take notes if needed
- Book on first aid (Fuerst n.d.)

You can get these items in pharmacy stores. Note that later, we will discuss alternative items you can use in lieu of some of

Family Survival Medicine Handbook

these in case you are caught off guard. The medicines are over-the-counter and thus do not require a prescription. Take them as prescribed on the packaging. However, the medicines mentioned are for a person with no underlying health conditions. Please be sure to consult your physician during the purchase of any medicine to determine whether or not you can take it, or if you should have the dosage adjusted, in case you have other health issues.

Now, let us learn how to make a DIY organized first-aid kit!

Initially, you need a water-resistant container. It is better if the box has multiple pouches so that you can categorize your items. Keep prescriptions in the container so that you have the dosage and warnings at hand. Regularly check the expiration date every 6 months and replace items as needed.

The items in the first-aid kit can be categorized as:

- Antiseptic: Wash your hands with soap and water before cleaning your wound with an antiseptic (like Betadine, a household chemical known to kill germs).
- Cream and ointments: Aloe vera gel can be used on burns, whereas anti-itch creams can be used for bug bites. Insect repellent creams can be applied to the skin to keep insects away.
- Medicines: Keep painkillers, cough medicines, antacids, and antidiarrheal medicines in separate compartments and mark them with a marker if necessary.
- Bandages and wraps

- Medical tools: Includes items like thermometers, masks, gloves, eye protection, scissors, tweezers, and flashlights.

When you are leaving for a wilderness experience or if you are disaster-stricken in your household, you can perform a risk assessment regarding the type of injuries you are more likely to encounter. Also, you can customize your first-aid box accordingly. For example, if you are leaving for mountain climbing or expecting potential tornadoes, you may require some extra bandages and anti-septic rather than a burn ointment or anti-diarrheal medicines.

Now that we've covered the basics, it's time to jump into action! In the coming chapters, you will learn in more detail about how to master handling different emergencies.

Takeaway

In any emergency, both mental and practical preparedness are equally important in determining your survival in the situation. You can start your mental preparedness today by attending seminars and events, holding drills, and practicing positive reinforcement. As the first step to practical preparedness, you can start by making your disaster preparation kit and a first-aid box.

3
TAKING CONTROL OF THE SITUATION: INITIAL ASSESSMENT

How Much Do You Know?

When a sudden injury occurs in a situation when you are cut off from the outside world, the first step is to survey the situation and the person. However, before beginning with step-by-step instructions on what to do, let us take a quick quiz to test your current knowledge.

1. In an emergency, who is the most important person?

    a. You
    b. Emergency medical services
    c. The injured person
    d. Bystanders

Answer: You are the most important person on the scene. Time is of great importance in managing any injury and you need to start acting promptly. Hence, you are the most efficient human resource available when disaster strikes.

2. What does CPR stand for?

    a. Cardio pediatric resuscitation
    b. Cardiopulmonary resuscitation
    c. Cardiopulmonary revival
    d. Cardiopulmonary recovery

Answer: Cardiopulmonary resuscitation is a life-saving procedure that is a combination of chest compressions and breathing assistance. CPR is only given to a person if they have stopped breathing, are breathing abnormally, or if their heartbeat has stopped.

3. At what rate should you aim to give chest compressions when giving CPR?

    a. 120-140/minute
    b. 100 /minute
    c. 100-120/minute
    d. 90-100/minute

Answer: Both adults and children should have their chests compressed at a pace of 100-120 compressions per minute. Chest compressions may be performed to the beat of Mark Ronson and Bruno Mars' hit song "Uptown Funk." If you haven't heard this one yet, other songs with the appropriate beat to follow are:

- "Stayin' Alive" by The Bee Gees
- "Dancing Queen" by ABBA
- "I Will Survive" by Gloria Gaynor
- "MMMBop" by Hanson
- "Girls Just Want to Have Fun" by Cyndi Lauper
- "Imperial March (Star Wars)" by John Williams
- "Baby Shark Dance" by Pinkfong (Chen 2017)

When giving CPR, thirty chest compressions are given, immediately followed by two rescue breaths. This makes up a single cycle of CPR (Lee 2012). If you are not confident in administering rescue breaths, you can give only chest compressions.

We will discuss the steps of CPR in detail in the later chapters of this book.

4. What is the first thing you should do if a victim isn't breathing normally?

    a. Check for a neck pulse
    b. Give thirty chest compressions
    c. Call 911
    d. Give two rescue breaths

Answer: If a victim isn't breathing normally (see pg. 45), you should first call for emergency medical help, if at all possible. You can then begin CPR. Your knowledge, ability to assess of the patient, and understanding of the protocol for treatment will get better as you read this chapter further.

5. How should you open the airway of an unconscious adult victim?

    a. Tilt the head sideways
    b. Tilt the head backward and gently lift the chin
    c. Move the tongue sideways with your fingers
    d. Tilt the head to the front

Answer: Use the head tilt-chin lift procedure on an unconscious adult victim to open the airway. Gently tilt the person's head back by resting your palm on his or her forehead. Then, using your other hand, push the chin forward to free the airway.

Head Tilt-Chin Lift

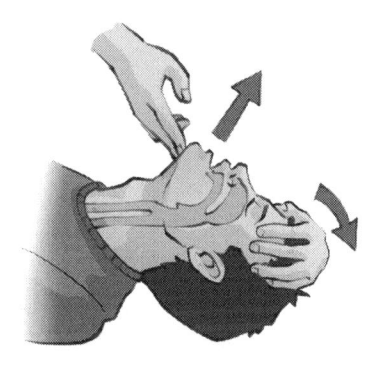

Now, let us further discuss the sequence of these procedures.

The ABCs of Survival Medicine

This next section is jam-packed full of extremely important information and we don't want you to get lost in the details of what's being explained. Here is a diagram leading you through what we will discuss as it relates to the next steps. If at any point you get confused about what to do next and where you are in the process, please use this chart to clarify. You can also take notes, as they can help you remember where you are in assisting the patient so that you know what to do next. You can refer to the SOAP method (see pg. 59) for more details.

Family Survival Medicine Handbook

Emergency in the wilderness → Survey the scene → Call EMS if severely injured → Obtain consent → Start with ABCDE

Suspected neck injury?

Airway and breathing
- Is the person breathing?
  - Yes → Distressed? / Choking? → If yes to both, refer to pg. 131
  - No → CPR (see pg. 116)

Circulation
- Check pulse
  - Yes → Is it too fast or slow? If yes, see pg. 90 and 92
  - No → CPR (see pg. 116)

Disability
- Grade by AVPU (see pg. 48)
- If unresponsive (U), assess ABCE again

Environmental exposure
- If skin is hot → Heat stress (see pg. 49, 71, 211, 262); Cramp (see pg. 211); Exhaustion (see pg. 211); Stroke (see pg. 212)
- If skin is cold → Hypothermia (see pg. 214)
- Anemia (see pg. 50)

After ABCDE, perform head-to-toe or SAMPLE → Check vitals → Transport the patient to a nearby health center if and when possible → If cannot transport, address individual injuries (see Chapter 4) until help arrives

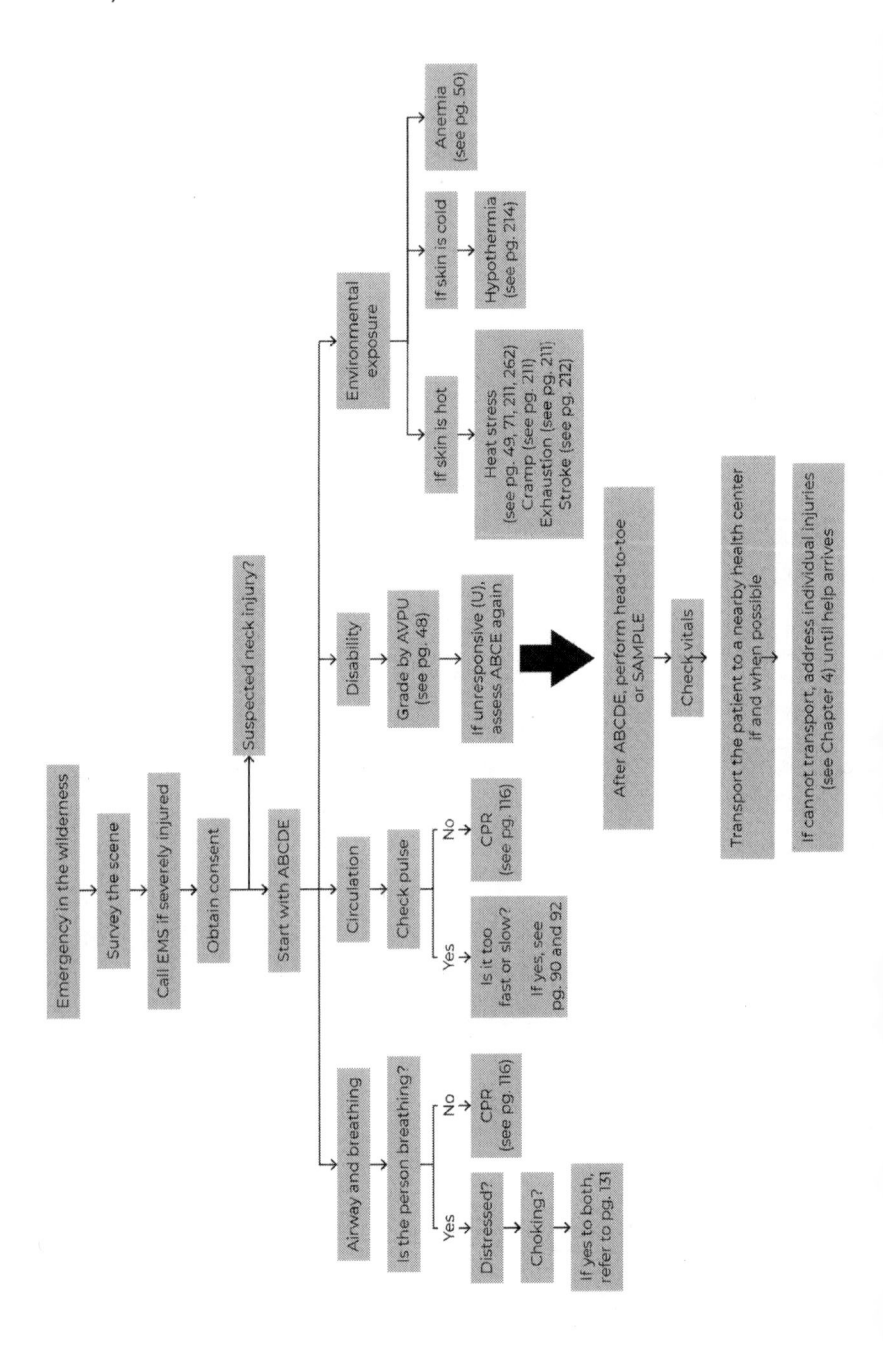

So, an emergency has just occurred. Maybe you are out camping and someone fell or maybe you are at home with no electricity or internet during the worst freeze yet. Suddenly, a family member or friend collapses. What is the first thing that you do?

When faced with a stressful situation, sometimes it's easy to forget the basics. That said, if you start to feel faint, then pause, ask for assistance (if possible), or do some deep breathing, and then go back to the situation at hand.

Step 1: Assess Scene Safety

Initially, your attention should be directed to surveying the scene of injury. Make sure that the cause of the injury is no longer hurting people, thus the scene is safe. If someone is injured during an earthquake and you are trying to help them but aftershocks continue to occur, you may want to look out for rock falls or any other structural collapse that could cause you harm or that may further injure your patient. You do not want another hurt person when you should be helping the victim.

Another part of scene safety is making sure you don't make your patient sick or vice versa. So, a good practice is to carry a mask with you so that in the event of an emergency, you can use it as a barrier to prevent the sharing of harmful germs. It is just as important to protect yourself from infection as it is to keep your patient safe.

Overall, determine scene safety and next steps by asking yourself these questions:

- Is the area safe? If the injury is still likely to occur again, then can I safely move the patient to a safer location without getting injured? Or can I maybe render help there but I just need to wait for something to clear, for example, the weather?
- How did the injured person get hurt?
- What is the seriousness of the situation? How sick or injured is the patient? If the situation seems bad enough, call EMS for help immediately.
- How many people are there who require your help? You may need to triage (categorize) the patients according to the seriousness of their injury—i.e., the most serious patients with a potential risk of sudden fainting need to be attended to immediately. A patient with moderate injuries, for example, a sore knee from a fall, is to be assisted next, and a patient with minor injuries, like a simple cut with minimal bleeding, can wait until you have attended to more serious ones.
- Are you protecting yourself from the environment and from hazards the victim might pose to you, such as by wearing gloves, a mask, and eye protection when coming in close contact with the patient? You have to limit the risk of infection from skin or environmental pathogens (germs) both for you and the patient (Healthywa.wa.gov.au n.d.).

Step 2: Consent

If the patient is conscious and can respond verbally, you must introduce yourself and ask for consent to transfer them to a safer place and to help them. Do not proceed if they say no. Respect the decision of the patient. If the patient is unconscious, it is regarded as implied consent and you can go forward to help the patient. If the patient cannot speak due to their injuries but makes gestures like pointing to their injured body part, it is understood that consent has been given and you can proceed further.

Special Note: Neck Injuries

If you suspect a neck injury, without moving the neck, gently feel along the spinous process, which is an easily felt projection in the middle of the back and neck. When you do so, a conscious patient can feel pain (point tenderness), which signifies that there is an injury to the backbone of the neck. If no pain is felt, it generally means there is no significant bone damage. If the neck is injured, or neck injury is suspected, neck movement is to be avoided as much as possible. Try your best not to move the neck much to open the airway if needed. But, if required, you can do this by performing the jaw thrust maneuver.

Jaw Thrust: If neck injury is suspected, open airway like this

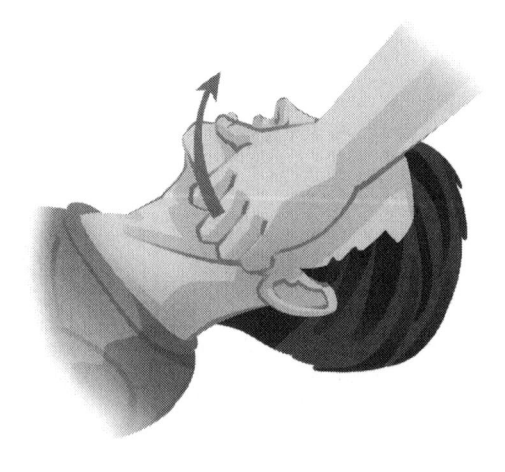

[Note: if your patient is unconscious (see pg. 41), skip ahead to pg. 36.]

If your patient is conscious (see pg. 41), keep their head, neck, and spine in a neutral position at all times and apply a splint if available (see the next illustration for an example of a neck splint and the neutral position).

Neck Splint

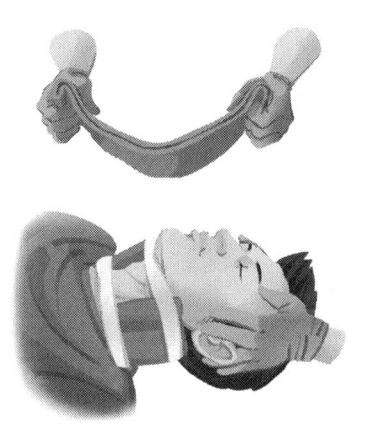

If the neck is at an abnormal angle, it should be straightened by gently placing your hands spread apart at the back of the head and pulling steadily and slowly on the head along the midline axis (the imaginary line in the middle of the body which divides the body into two halves, right and left, lengthwise). Move the neck to the neutral position in line with the spine and keep it there (see the next illustration—note that it shows the neutral position as well as how to hold the neck to move it to the neutral position). This helps in protecting the spinal cord so that you do not paralyze the patient (Healthywa. wa.gov.au n.d.; Clinical Quality & Patient Safety Unit 2019). Note that it will help you maintain position if your elbows are on your knees or the ground to help you support your patient's head and neck. You can also apply a neck collar if available or tuck pads on either side of the neck to prevent its movement at rest. As best as possible, keep the patient in this position until help arrives.

Manipulation of the Neck

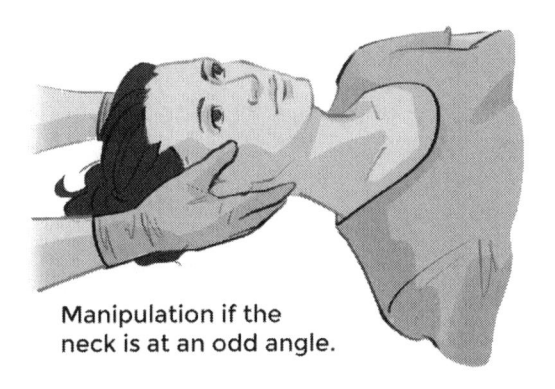

Manipulation if the neck is at an odd angle.

If a patient is unconscious (see pg. 41) and has sustained an injury to the head or neck, treat as if the neck is fractured. After trying your best not to move the patient while stabilizing them in terms of ABCDE (airway, breathing, circulation, disability, environment; see pg. 40), until help arrives, place the patient in the recovery position to maintain support of the neck, and, if available, a splint (rigid material used for supporting a broken bone) should be used to prevent movement (Healthywa. wa.gov.au n.d.). Note that this also helps to maintain their open airway. To move the patient into the recovery position, do this:

1. Bend down near your patient and grasp their opposing hand.
2. Move it above their head.
3. Move the hand closest to you so that it is supporting their neck.

Taking Control of the Situation: Initial Assessment

4. Grasp the knee and elbow of the same side that is supporting the neck.
5. Roll the patient onto their side.
6. Now the patient is in the recovery position.

The patient's top arm should support the head and neck area, and the bent knee will prevent the patient from rolling too much (Nhsinform.scot 2022; Bennett 2019). See the illustrations below.

Recovery Position Steps

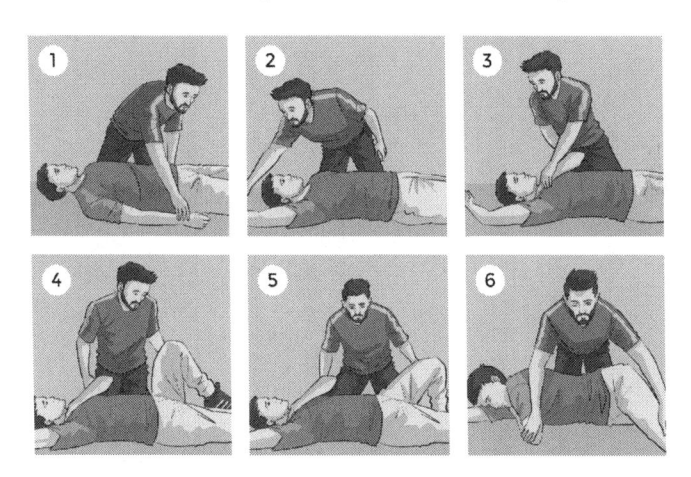

Recovery Position
To keep the airway open and clear until help arrives

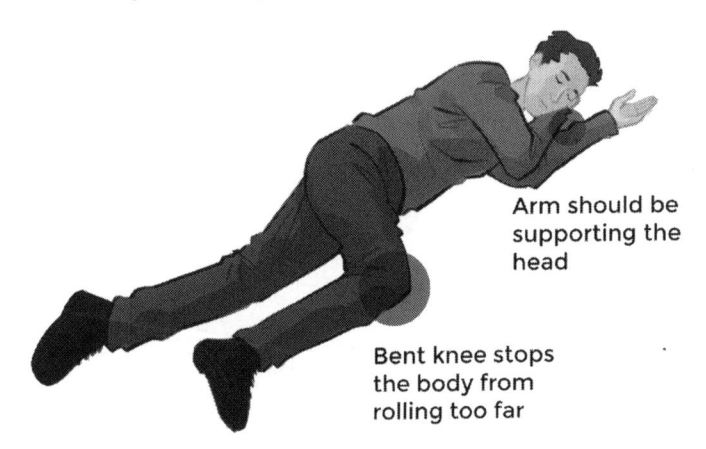

Arm should be supporting the head

Bent knee stops the body from rolling too far

Neck Splint

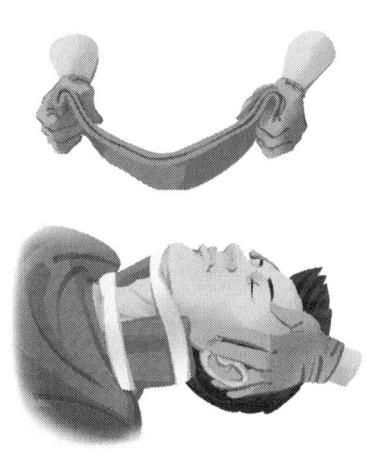

Step 3: Severe Injury Assessment

This section will cover how to assess the patient in the event you do not know if they can survive their injury or illness.

When a sudden injury occurs, after surveying the scene, if the injury is serious and if possible, call EMS immediately. Then, start by putting on your PPE (personal protective equipment) and getting consent. Check for a neck injury. Following this, the next step is a systematic approach for the immediate assessment of a patient. It is a standard approach followed very often by experts in emergency medicine known as the ABCDE approach (Thim et al. 2012). ABCDE stands for airway, breathing, circulation, disability, and exposure. We will be using this method to guide you through the management of the patient.

Airway and Breathing

Can air freely flow into and out of the body without being blocked, and is the body performing that operation? The body needs the ability to receive and release air because when air is inhaled, it contains oxygen, which energizes the body, and upon exhalation, it allows the body to get rid of waste or carbon dioxide. This very important transportation of air happens by way of your windpipe (trachea, airway), which you can feel the rings of at the front portion of your throat as it goes down and ends inside your chest via your lungs. See the illustration below.

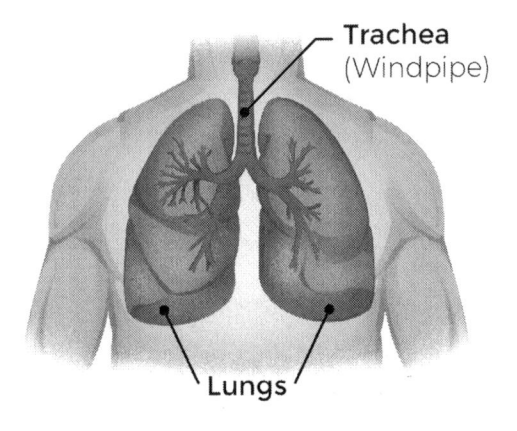

This act of receiving oxygen and getting rid of waste or carbon dioxide by way of our trachea is called breathing.

For us to live, we must be able to breathe. Without oxygen, our organs cannot function and our bodies shut down. If your patient is conscious, it means he or she is breathing.

Taking Control of the Situation: Initial Assessment

Hence, the next step is to determine your patient's consciousness, which ties into determining if your patient's airway is open and they are in fact breathing. If a person can talk, or respond by any other non-verbal means, their airway is open and working, which means they can breathe. To establish if your patient is conscious, start by asking them questions such as "Are you okay?" or "Can you hear me?" If there is no response to your verbal communication, make gentle physical contact by touching the victim's shoulder and repeating your questions. Sometimes an injured person may not respond to gentle contact. If so, try applying a painful stimulus, such as pinching the back of the patient's arm. If the person responds by making eye contact or some other gesture, they are conscious. If there's still no response, consider them unconscious.

For an unconscious patient, the next step is the airway assessment. Look for an open airway by watching the rise and fall of their chest. If the breathing pattern is abnormal and you suspect that it is due to breathing complications, begin CPR. If not, look for any other causes, like anxiety, which can be taken care of by following steps for mental preparation (see pg. 16), and if the cause is unknown, see pg. 45 for discussion on abnormal breathing. If you do not see chest movement, it suggests that the airway is closed and something may be blocking it. Hence, you should focus on removing the agent and clearing the airway. The blockage may either be a clot, debris, or objects in the mouth. Open the mouth to check if you see something, and if yes, remove it. The next illustration will guide you.

41

Family Survival Medicine Handbook

Removing Foreign Body from the Airway

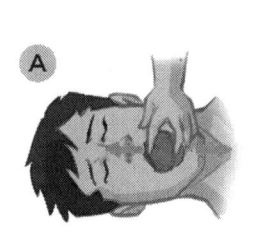

A Open the mouth of the injured

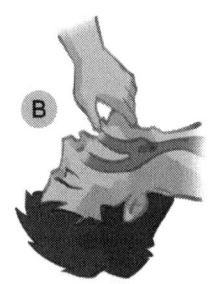

B Perform jaw lift maneuver (discussed in Chapter 5)

C Locate the foreign body

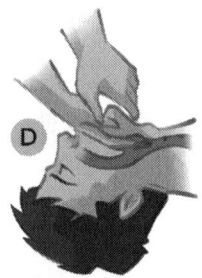

D Insert your finger as shown

E Sweep your finger and "hook" the object

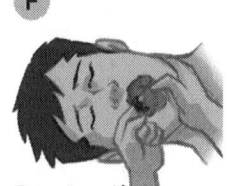

F Remove the foreign body

If you do not see anything, another consideration in an unconscious patient is airway obstruction due to the tongue falling back into the throat. Perform the head tilt-chin lift technique to investigate. Place one hand on the patient's forehead and gently tilt their head back while elevating the tip of their chin with two fingers. Then, using the index (pointer) and middle fingers, physically push the lower jaw upwards where it links with the back. See the illustration below.

Head Tilt-Chin Lift

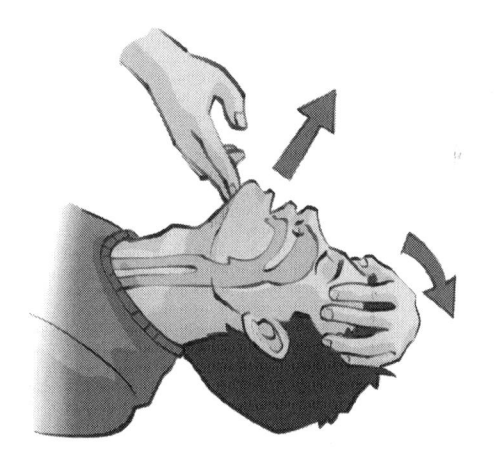

This maneuver helps in moving the tongue away from the back of the throat (Thim et al. 2012). Ultimately, you know that you have cleared your unconscious patient's airway and it is open when you find signs of breathing, like the presence of breathing sounds, the rise and fall of the chest, or the feeling of air leaving the nose or mouth.

B reathing?

After checking to see if your patient is breathing using all the just mentioned techniques, including opening their airway, if you determine that your patient is not breathing, immediately begin CPR (see pg. 116). If your patient is breathing abnormally due to respiratory failure, begin CPR. For more information on abnormal breathing, see pg. 45.

If you determine that the patient has an open airway, is conscious, and is breathing, proceed further to head-to-toe examination (see pg. 59) or SAMPLE (see pg. 64). If the patient is unconscious but has an open airway and is breathing normally, transfer the patient to a nearby medical facility as soon as possible to look for other causes of unconsciousness and for assessment of any brain injuries. Until help arrives, proceed to do a head-to-toe examination (see pg. 59) and treat accordingly, continuously check their vitals (see pg. 65), talk to the patient, loosen anything that is tight and maybe keeping them from continuing to breathe easily, and place them in the recovery position (see pg. 36) to help keep their airway open (Hatraining.com n.d.).

If your patient is conscious and experiencing distress or difficulty breathing, ask, "Are you choking?"

Choking (object in the airway)

Choking is described as a partial or total blockage of the airway caused by a foreign body (e.g., food, a bead, a toy, etc.). A victim who is choking cannot talk but may make a high-pitched sound during attempts to breathe. He or she will rapidly become bluish in color and unconscious if the blockage is total. This is a life-threatening situation and needs to be managed immediately (Stöppler 2021).

If your patient is choking, see pg. 131.

Abnormal Breathing

Abnormal breathing can occur due to anxiety, choking on a foreign body, respiratory system failure (or via the lungs failing to get the body enough oxygen and release waste), and swelling in the anterior neck. Most instances are caused by heart and/or lung issues. Causes like anxiety can be managed in the household or wilderness by following steps for mental response, which include remaining calm yourself, helping your patient remember to breathe, giving them an object to touch, reminding them that you are present, and asking what you can do to help (Healthline.com 2021). Choking can be managed by the life-saving maneuvers discussed on pg. 131. However, respiratory system failure and neck swelling are more serious and require professional assistance. The abnormal breathing sounds are to be identified to the best of your judgment. If abnormal breathing is complicated by respiratory failure, or there is suddenly no breathing or pulse (see pg. 40, 51), CPR is your next action step (see pg. 116).

Circulation

After having confirmed that your patient's airway and breathing are fine, check their heart. Circulation refers to your heart's ability to pump blood throughout your body. It is important because, when healthy, our heart pumps enough blood throughout our body to give it the necessary fuel to carry out basic life functions like breathing, breaking down food to give us energy, maintaining our body's optimal internal temperature for functioning, ridding the body of waste, and so on. When this system is disrupted (i.e., through severe bleeding or if the heart stops pumping), it can mean that the body is not able to carry out those basic life functions. That's why it's important to see if your patient's heart is beating. You can learn this by checking their pulse (see pg. 51).

Furthermore, if a person can talk, their heart is beating. If your patient's heart has stopped, immediately begin CPR (see pg.

116). If you suspect your patient's heart is beating too slowly or fast, see pgs. 90 and 92.

Examine for substantial blood loss. Survey the ground for obvious blood drippings or feel for it with your hands. Also, check under thick clothes for hidden blood loss and put your palm under the patient to ensure that blood is not spilling onto the ground or snow undetected. If there is visible, severe blood loss, apply direct pressure and/or use a tourniquet. A tourniquet is a cord or tight bandage for stopping the flow of blood and is thus tied just above the wound. It is applied around your arms or legs. However, if you have a bleeding wound from your head or abdomen, tying up a tourniquet is not an option. You have to go for other measures like applying direct pressure to stop the bleeding. If your patient is losing a lot of blood, see pg. 169.

Disability

Although a basic evaluation of consciousness has already been carried out, once the airway, breathing, and circulation have been addressed, it is time for a more in-depth consciousness evaluation.

In order to assess the patient's level of consciousness, ask specific questions, like, "Do you know what time it is? Is it day or night? Do you know where you are at the moment? Do you know who you are/who I am?" This tells us the patient's level of brain functioning. We can then grade the patient based upon AVPU:

- Alert (A): Oriented to time, place, and person
- Voice responsive (V): Utters sounds or words when spoken to
- Pain responsive (P): Moves their body or utters sounds in response to a painful stimulus such as pinching the back of their arm or briskly rubbing the bony center of their chest with your knuckle
- Unresponsive (U): The patient is unresponsive to all stimuli (Thim et al. 2012)

If you find your patient unconscious—i.e., unresponsive (U) to all stimuli, like voice or pain—check ABC to see what needs to be addressed. If your patient cannot talk but is responsive to other stimuli, you can deem the patient conscious and move on to further assess other parameters.

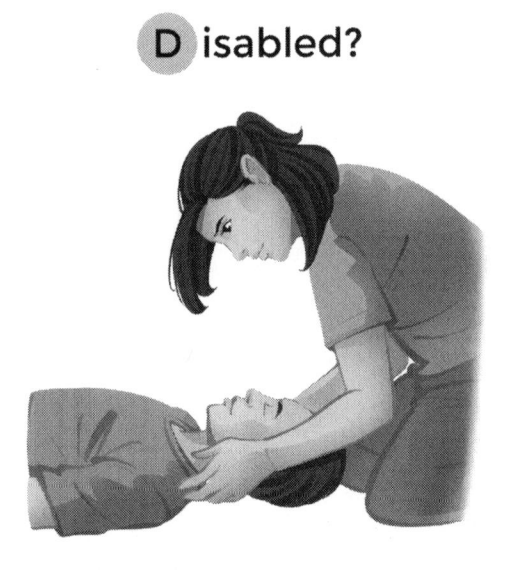

Disabled?

Also, can your patient move under their own power or do they need help? Your patient's inability to move could be suggestive of a neck injury. If you suspect a neck injury, see pg. 33.

Environmental Exposure

Once ABCD checks have been addressed, then check for environmental exposure. Environmental exposure refers to the clues that explain the patient's condition (Thim et al. 2012). These may include signs of trauma, bleeding, skin reactions (rashes), or needle marks. What do you see that can help you determine what caused your patient's injuries?

While doing your best to respect your patient's privacy and dignity, along with their neck safety, clothing should be removed (ideally with scissors if you have them) to allow a thorough physical examination to be performed. To get an estimate of the body temperature, feel the skin or use a thermometer when available.

Signs to keep in mind when touching the skin are:

- Heat: Examine your patient. If your patient is too hot, it could be a sign of heat stress (see pg. 211). Heat stress symptoms like cramps (pg. 211), exhaustion (pg. 211), and stroke (pg. 211) will be discussed in later chapters of the book. If the skin is hot, it is good to take the temperature to know if the internal body temperature is too hot (Centers for Disease Control and Prevention n.d.).

- Cold: Is your patient too cold? This could be a sign of hypothermia (see pg. 214). Hypothermia is the decrease in normal body temperature due to prolonged exposure to cold. Hence, it would be good to check your patient's internal body temperature with a thermometer. For how to take the temperature with a thermometer, please refer to pg. 70.
- Anemia: This is defined as the condition when your body does not have enough healthy red blood cells. They are important because they carry oxygen to the tissues in your body, which in turn gives them the needed energy to function properly (Mayo Clinic Staff 2022). Here is how you can identify anemia. The palms of the hand appear pale with no reddish hue. The layer that covers the inner surface of the lower eyelid (see next illustration) is paler than normal. You can also look for the capillary refill time. To determine this, press the tip of your patient's finger (on the non-fingernail side) with your thumb, then look at the time taken for the restoration of red color to their skin. The normal capillary refill time is usually <=2 seconds. Patients with anemia can present with lower body temperature than normal. For the thermometer section and temperature measurement, see pg. 70.

The Palpebral Conjunctiva: a site to look for anemia

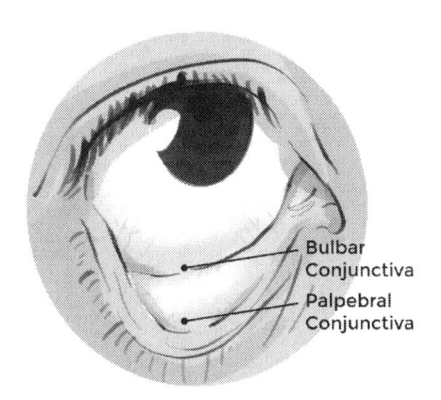

- Bulbar Conjunctiva
- Palpebral Conjunctiva

Pulse

Each time your heart contracts, it causes your artery to pulsate or beat. To find the pulse, press on the skin lightly where there's a large artery (the tube-like structure that carries blood throughout your body), right below the skin. An artery can be found on the side of the neck, inside the wrist (just below the thumb), or on the shoulder of an infant. Pulse is felt when you find a small beat under the skin. See pg. 66 for normal pulse measurements.

Now we will look at the sites and the different methods of measuring the pulse in our bodies (Healthline 2019).

How do you take a pulse?

Your neck pulse is the beating felt on the carotid artery. Place your two fingers (pointer and middle fingers) on the side of the windpipe (trachea) just below the jawbone. Once you feel the pulse, count the beats up to 1 minute, or up to 15 seconds and then multiply by 4. The number you calculate equates to the pulse measurement.

(Note: do not check the left and right carotid pulse at the same time, as it can sometimes lead to a sudden drop of blood pressure in the patient.)

Circulation?
Neck Pulse Check

The thumping felt on the radial artery is your wrist pulse. Place your index (forefinger/pointer finger) and middle fingers directly below the thumb on the inside of your opposing wrist. After that, count the beats.

(Note: do not use your thumb to feel the pulse, as the artery of the thumb can interfere with the wrist pulse.)

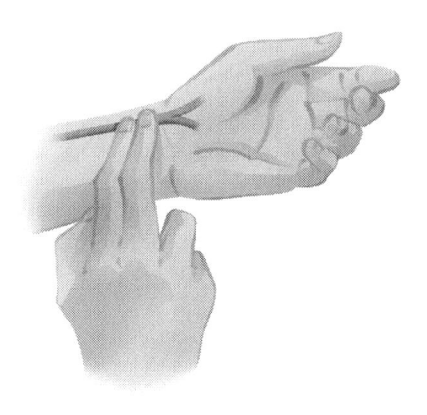

C irculation?
Wrist Pulse Check

Your elbow pulse is the beating felt on the brachial artery. It can be found by putting your pointer and middle fingers on the inner aspect (2 cm) from your bicep's tendon (stiff part of muscle where it connects to bone) and 2-3 cm above the elbow joint.

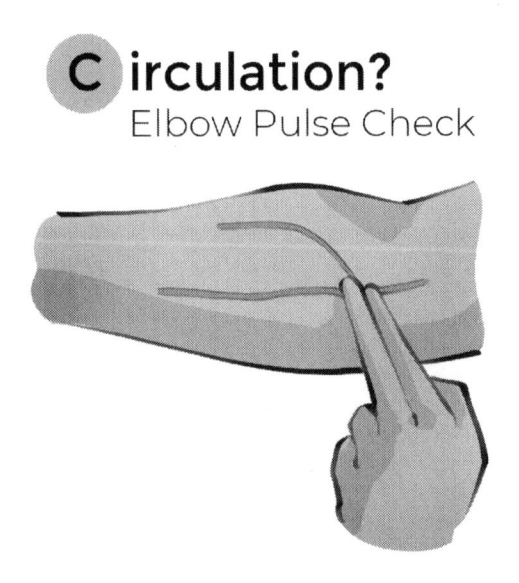

Circulation?
Elbow Pulse Check

Femoral artery beating indicates the femoral pulse. It is felt on the inner aspect of the thigh, up approximately midway.

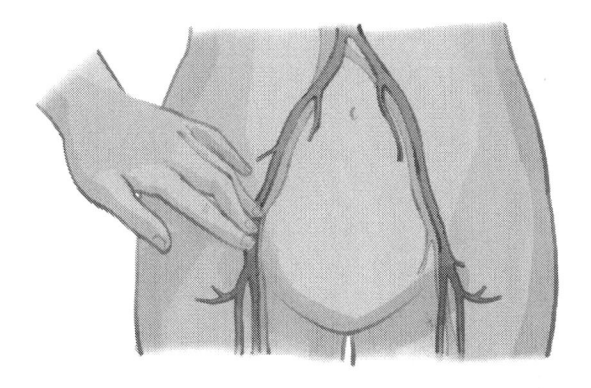

C irculation?
Groin Pulse Check

Popliteal pulse is the beating of the popliteal artery. It is felt in the soft spot behind the knee.

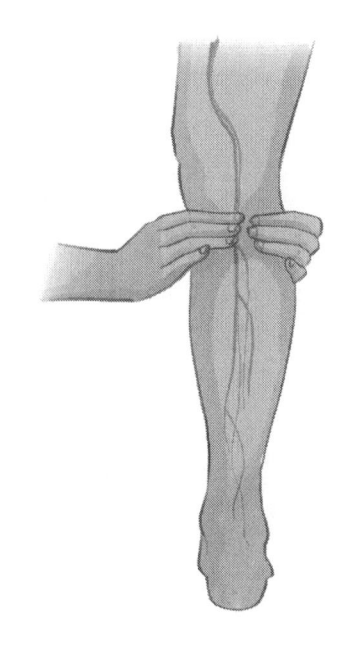

Circulation?
Back of Knee Pulse Check

Taking Control of the Situation: Initial Assessment

How do you check infants' and children's pulse?

Infant (birth to 1 year) : The upper arm, commonly known as the brachial pulse, is the best place to feel an infant's pulse. Place the infant on their back with one arm bent and one hand up by the ear. Feel for the pulse on the inner arm between the shoulder and the elbow with your index (pointer) and middle fingers. Then count your pulse.

How to Check Baby Pulse

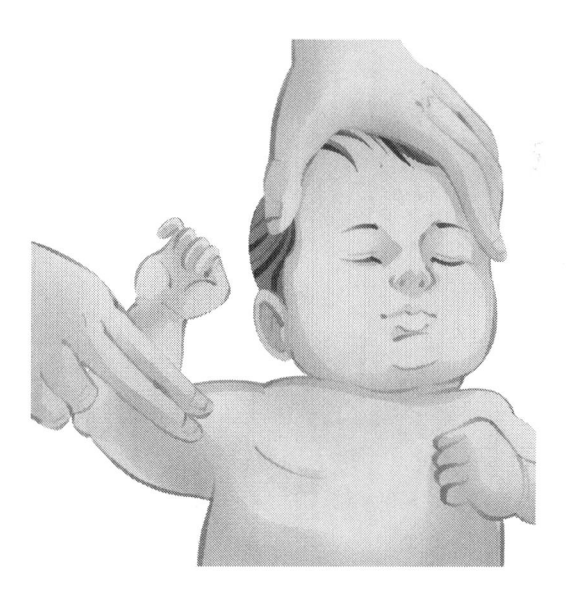

Children's pulse (1-5 years of age) : In children, you can either take the radial (wrist) pulse or carotid (neck) pulse in the same way as in adults (see pg. 52).

57

Special Note

Since this overall evaluation is so important, let's go through some facts worth repeating. While going through ABCDE, as soon as you realize your patient is not breathing or is breathing abnormally (due to respiratory problems) or has no pulse (see pg. 52), begin CPR immediately (see pg. 116) (Dictionary.com n.d.; Merriam-webster.com n.d.).

The medical term used when there is no neck pulse is cardiac arrest. Cardiac arrest means that your patient's heart has stopped performing its job of pumping blood throughout their body (Mayo Clinic Staff n.d.). And as you know, your body needs blood to have the energy to perform many functions, including such things as breathing. We learn whether a person's heart is functioning properly by checking their pulse, as discussed on pg. 52 in detail (Dowshen 2018).

If multiple people are severely injured, who do you treat first?

Which patient is dealing with more ABCDE issues? That determines who you treat first. Ensure that the place you treat your patient is separate from climate hazards, clean, and with restricted access to hopefully prevent contamination.

Also, during this time, you should take notes on each patient via the SOAP method (Subjective, Objective, Assessment, and Plan) so that others can help you and you know how far your patient has come since treatment began:

S: Subjective. Refers to the patient's complaints in their words. For example, the patient may complain of pain.

O: Objective. Refers to your judgment. For example, if the patient complains of pain, you assess whether the part being spoken of is tender to touch.

A: Assessment. What could be the cause?

P: Plan. According to your assessment, how do you proceed with treatment? This determines your plan.

Diagnostic Techniques

We have now stabilized the patient, thus assessed ABCDE. Now, let's move on to diagnosing the injury. For this, a complete head-to-toe examination is to be done. This sequence should be practiced from beginning to end such that none of the injuries are missed. You can practice this sequence on others as well as on yourself.

In what is known as the DCAP-BTLS examination, we assess for deformity (abnormal size or shape), contusions (bruises), abrasions (scrapes), punctures (holes caused by trauma), burns, tenderness (or soreness), lacerations (cuts), and swelling (EMTResource.com 2014).

This sequence of examinations should only be interrupted to provide life-saving care such as CPR or to stop life-threatening bleeding. Always wear gloves when performing ABCDE and head-to-toe examination. It helps prevent soiling of your hands and also prevents the exchange of disease between you and the patient.

The general principles of the focused physical assessment are as follows:

1. Start assessment of the patient from the top of their body and work your way down.
2. Try not to aggravate the existing injuries while looking for others. Hence, move the patient as little as possible.
3. When examining the patient, maintain constant communication even if he or she seems unconscious. It will prepare the patient for the upcoming actions and decisions.
4. Look for damage and even cut off the clothing of the patient, if necessary, when injuries are suspected.
5. Ask about pain, discomfort, numbness, and abnormal sensations constantly when examining the patient.
6. Feel all the relevant body parts gently for any abnormalities (Forgey 2020).

Now that we have learned the basic principles, let us begin with the examination sequence:

- Head: Examine the area for lumps, swelling, injury, discoloration, and abnormalities. Blood or fluid leaking from the ears, nose, or mouth may indicate a skull fracture that requires rapid expert attention. Inquire

Taking Control of the Situation: Initial Assessment

about any loss of consciousness, discomfort, or other unusual symptoms. For:

o   Loss of consciousness, see pg. 41, 76
o   Headache, see pg. 147
o   Ear trauma, see pg. 84
o   Eye trauma, see pg. 81
o   Nose trauma, see pg. 86

- Neck: Look for noticeable damage or irregularity of the windpipe (trachea; see illustration on pg. 40). Ask about pain, discomfort, and difficulty breathing. Feel along the vertebrae (spine ridges in the midline at the neck and back) for pain (Forgey 2020).
- Chest: Look for swelling, unusual cracking sensation with slight pressure, or pain. Feel for instability. For:

Chest trauma, see pg. 92
Difficulty breathing, see pg. 45, 116

- Abdomen: Gently press on the abdomen with your hands spread wide. Ask the patient to bend their legs if possible. This will aid in better evaluation of the patient. Look for stiffness of the abdominal muscles, muscle spasms, or expansion of the abdomen. Ask about pain and discomfort during examination. For abdominal pain, see pg. 94.
- Back: Slide your hand under the patient. Look for any active bleeding. Feel along the backbone for any tenderness. For spine trauma, see pg. 33.

- **Pelvis/Hip:** Place both of your hands on the top front of the hipbone on both sides simultaneously, pressing gently down and pushing towards the center of the body. Ask about pain and feel for instability. For hip or pelvic pain, see pg. 97.
- **Legs:** Examine one leg at a time. Place your hands around the leg and run your hands down from the groin (area of your hip between your stomach and thigh; see next illustration) down to the toes, squeezing as you go (Forgey 2020). Note if there are deformities or lack of circulation, sensation, or motion in the toes. Also, check for resistance by asking the patient to move their legs in the opposite direction of the force you are applying. Check for the movement of the foot. Repeat for the other leg. For bone injury, see DIY splints on pg. 106.

Groin Area

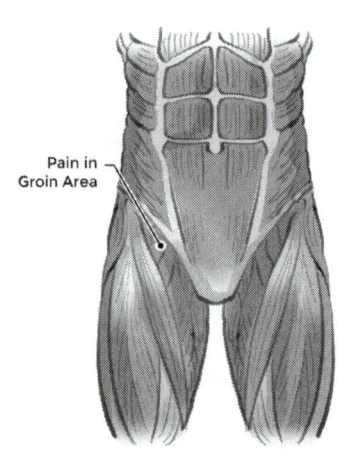

Pain in Groin Area

- Shoulders and Arms: Examine one arm at a time. With hands placed wide, squeeze the shoulder and run your hands down to the arms and to the fingers. Keep squeezing throughout (Forgey 2020). Check for deformities or lack of circulation (elbow pulse), sensation, and motion in the fingers. Check for muscle power by asking the patient to grasp your index finger with their hands. Look for their grip power. Repeat on the other shoulder.

Decision chart:

- If head injury with:
  o Loss of consciousness, see pg. 41,76
  o Headache, see pg. 147
  o Ear trauma, see pg. 84
  o Eye trauma, see pg. 81
  o Nose trauma, see pg. 86
- If chest injury due to trauma, see pg. 92; for difficulty breathing, see pg. 45
- If abdominal pain, see pg. 94
- If hip or pelvic pain, see pg. 97
- If suspected bone injury, see DIY splints on pg. 106
- If joint trauma, see pg. 102
- If suspected broken bone, see DIY splints on pg. 106

SAMPLE and Checking Vital Signs

If you believe your patient is suffering from a medical issue due to long-term illness and not trauma (sudden injury), then instead of the head-to-toe exam, you will have to go for a SAMPLE assessment. Hence, after you have stabilized the patient by ABCDE, move on to SAMPLE or head-to-toe examination, and then move on to checking their vital parameters. SAMPLE gives us an idea of the questions we need to ask the patient.

SAMPLE stands for:

- Signs and symptoms: Signs are what we observe while examining a patient and symptoms refer to what the patient is experiencing. For example, if a patient is having difficulty breathing, it is a symptom, whereas if, on examination, we find that the patient is breathing rapidly, it is a sign.
- Allergies: Ask the patient if they have any known allergies.
- Medications: Ask the patient if they are currently on any medications. It is necessary to know their history of medicines to evaluate the drug reactions and the underlying disease conditions.
- Past medical history: Ask the patient about their past medical history and what medicines they are taking for it.
- Last oral intake: Ask the patient when they last ate, drank, or took medication.
- Events leading to the cause: Ask the patient about

the causative factor for the underlying condition (Coyne 2021).

Asking these questions will help in establishing the underlying cause of the present condition of the patient.

Vital Signs

The vital signs include heart rate, measured by taking the pulse, breathing rate, blood pressure (circulation), and body temperature. All of these signs signify that the most important systems in our body are working well. They also help us diagnose any underlying conditions.

In an emergency, start with the ABCDE. Then, go for a head-to-toe or SAMPLE examination, which depends on the injury of the patient. Then, check all the vital signs (pulse, blood pressure, temperature, and breathing rate) at frequent, regular intervals.

All of these measurements have a normal range, and values above or below them are considered abnormal. The normal ranges differ according to age and slightly according to gender. However, there may be a slight variation in the normal range in some individuals. It is extremely important to relate the clinical symptoms of the patient with the values obtained.

The normal range of vital parameters is given in the table below (Coyne 2021):

Age	Pulse (per min)	Respiratory Rate (per min)	Blood Pressure (mm Hg)	Temperature (degrees C and F)
Newborn (0 to 1 month)	100 to 205 bpm	30 to 60 bpm	Systolic: 60 to 76 Diastolic: 16 to 36	97°F to 99°F 36.1°C to 37.2°C
Infant (1 month to 1 year)	100 to 180 bpm	30 to 53 bpm	S: 72 to 104 D: 37 to 56	97°F to 99°F 36.1°C to 37.2°C
Toddler (1 to 2 years)	98 to 140 bpm	22 to 37 bpm	S: 86 to 106 D: 42 to 63	97°F to 99°F 36.1°C to 37.2°C
Preschool (3 to 5 years)	80 to 120 bpm	20 to 28 bpm	S: 89 to 112 D: 46 to 72	97°F to 99°F 36.1°C to 37.2°C
School-age (6 to 9 years)	75 to 118 bpm	18 to 25 bpm	S: 97 to 115 D: 57 to 76	97°F to 99°F 36.1°C to 37.2°C
Preadolescent (10 to 12 years)	60 to 100 bpm	12 to 20 bpm	S: 102 to 120 D: 61 to 80	97°F to 99°F 36.1°C to 37.2°C

Taking Control of the Situation: Initial Assessment

Age	Pulse (per min)	Respiratory Rate (per min)	Blood Pressure (mm Hg)	Temperature (degrees C and F)
Adolescent (12 to 15 years)	60 to 100 bpm	12 to 20 bpm	S: 110 to 131 D: 64 to 83	97°F to 99°F 36.1°C to 37.2°C
Adults (16 and up)	50 to 100 bpm	12 to 20 bpm	S: 90 to 120 D: 60 to 80	97°F to 99°F 36.1°C to 37.2°C

Take a patient's vitals according to the seriousness of the injury. In the emergency stage, if feasible, they should be taken every 5 minutes. After stabilization, they can be taken every 4 to 6 hours.

- Pulse: As previously discussed, measuring the pulse is an indirect measure of the heartbeat, as the pulse felt correlates to the heartbeat. To learn to measure the pulse rate in an adult, child, and infant, see pgs. 52-57. The pulse is generally most readily felt in the neck or the wrist. Other places where the pulse is felt are the front of the elbow (brachial pulse), the groin (femoral pulse), and the back of the knee (popliteal pulse). Not only is the rate of the pulse important, but it is also important to determine the rhythm of the pulse. An irregular pulse (one that is not beating at regular intervals) signifies an abnormality in heart rhythm. Deformed fractures (fractures which are altered from

normal shape) can injure the blood vessels and cause a decreased pulse.
- Breathing (respiratory) rate: You can measure the breathing rate of a person by watching the rise and fall of their chest. It is better to obtain this information when a person is lying on their back (supine) as you can also feel the rise and fall of their chest with your hand. It is better to distract the person when taking their breathing rate, because when a person becomes conscious that they are being watched, their breath tends to become shallow and quicker. For breathing difficulties, see pg. 45.
- Blood pressure: The force that blood flow exerts on a blood vessel's wall is known as blood pressure (Forgey 2020). Systolic pressure is the pressure exerted when the heart squeezes at its maximum capacity. Diastolic pressure, in contrast, is the pressure just before the contraction of the heart. Mean Arterial Pressure (MAP) is calculated using systolic and diastolic blood pressure and is the average pressure of blood received by the vital organs of our body. It is important to know the MAP to evaluate how well the blood flows through your body. The conventional measurement of blood pressure requires a cuff (sphygmomanometer) and a stethoscope (an instrument used to listen to internal body sounds). Have the patient seated and relaxed with the left arm at the level of the heart. The cuff is kept on the left arm and the stethoscope is placed where the brachial artery (elbow) pulse is felt. The brachial artery passes through the front side of the elbow and is the

major blood supply to the hand. The cuff is inflated to above the normal range and the sound of the pulse is heard in the stethoscope. Note when the sound appears and when it disappears. This is your systolic and diastolic pulse, respectively.

Sphygmomanometer

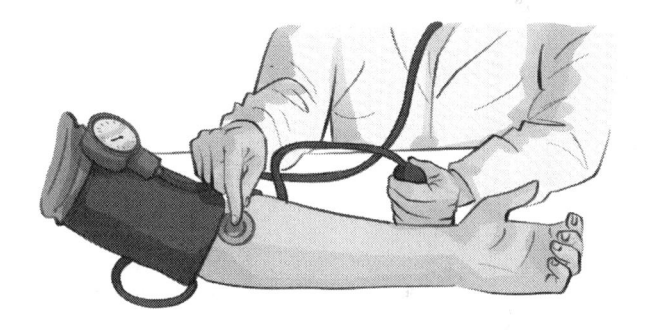

- Nowadays, there are several digital blood pressure machines available on the market, though they are known to be less accurate. These can be of help if you are not confident enough to measure blood pressure conventionally. If you do not have a cuff or digital blood pressure measuring set in an outdoor emergency, here are other ways to estimate blood pressure.
- If you can feel the wrist pulse, it means that the top (systolic) pressure is probably at least 80 mm Hg. If you can only feel the groin (femoral) pulse, the pressure is more than 70 mm Hg. When only the neck (carotid) pulse is felt, the systolic pressure is probably

at least 60 mm Hg (Forgey 2020). An increased pulse rate with low blood pressure is an indication of shock (see pg. 166).

- Temperature: Thermometers are used to measure the internal temperature of the body. Thermometers can be mercury, digital, infrared, or stick-on disposable thermometers. Mercury thermometers are the conventional ones but are no longer preferred due to the risk of breakage and toxicity of mercury. Digital thermometers are now widely used for measuring temperature. They are inserted either in the mouth, armpit, groin, or rectum. Infrared thermometers require no contact with the patient and are being widely used in this era of COVID-19. Stick-on thermometers have strips that change color on use. The strips are stuck to the forehead. They are single-use and prevent infection transmission. For fever and its management, see pg. 154. You can estimate the body temperature of a patient with a fever if the person's normal resting pulse rate is known. Each higher degree Fahrenheit will generally result in a pulse increase of 10 beats per minute. However, cases like typhoid fever are an exception, when there is a relatively slow heart rate for a high fever (Coyne 2021).

For breathing difficulties, see pg. 45
For shock, see pg. 166
For fever and its management, see pg. 154

After a thorough examination of the patient and their vital signs, you should know what the problem is and you can prepare to decide on treatment. The upcoming chapters of this book will provide an outline of the symptoms specific to the these injuries, along with others and their treatment.

Can You Make the Situation Safer?

Apart from the ABCDE of basic survival skills, there are certain measures that can be taken to make the situation safer and the patient more stable. These assessments assure that further damage is not done and the patient does not revert back to needing ABCDE measures again. These are as follows:

- Control the spine: This is already mentioned on pg. 47 (D for disability). Please refer to this for more details.
- RICE: This stands for rest, ice, compression, elevation. After a limb (an arm or leg) injury, you can promote healing and reduce swelling and pain with RICE (see pg. 103) (Healthwise Staff 2020).
- Direct pressure on a bleed: If a person sustains an injury that is bleeding profusely, apply direct pressure to it, preferably with thick padding. Ideally, the material to be used is a sterile gauze pad, but you may have to use clothes as padding in the wilderness. Maintain a steady pressure until the bleeding stops, then pack the wound by placing more gauze and topical ointment (ointment applied to a particular place, usually the skin) on the wound to help it heal. Use gloves when dealing with bleeding wounds to prevent blood-borne infection transmission.
- Heat exhaustion: This is the life-threatening rise in the

core body temperature of the patient. It occurs due to heat stress and decreased water content in the body due to heat. If this occurs, move the patient into the shade and give them water and fluids. For more information on patient management with heat exhaustion, see pg. 211.
- Hypothermia: This is the dropping of the body temperature below the normal range so that the vital systems of the body slow down. It is necessary to immediately manage the patient, as hypothermia can lead to heart failure and death. The patient should be kept warm with blankets as quickly as possible and given hot fluids. For a detailed description of the management of a patient with hypothermia, see pg. 214.

Principles of Survival Medicine

Now that we have gone into detail on patient diagnosis and assessment, let's again go back to the basic principles of survival medicine.

These principles are what we need to remember when we are applying survival aid in the wild or when the unthinkable happens and help is not on the way:

- Stay calm: When you have an injured person in the wilderness, you have a lot to do. But don't panic. Stay calm. Take deep breaths by slowly taking in air and letting it go. Know that you are the only help avail-

able to the patient. After you have settled down, get into action.
- Be prepared: Know the sequence of patient assessments and be prepared with a medical kit anytime you are going into the wild. You should take first-aid and wilderness medicine courses and also read and collect books like this one so that you feel confident and can promptly help the patient.
- Know your supplies/assessment: When helping the patient, know your supplies and resources. If available supplies are insufficient, and you can use the surrounding natural resources, do it! Save resources when you can so that they will last for a longer period, hopefully until help arrives (Thim et al. 2012).

Alertness and the ability to diagnose emergencies are the key factors in rescuing victims. You are the decision-maker in the scenario for the patient, so stay calm and think clearly so that you avoid making hurried decisions with a lack of forethought. Your calmness, agility, and preparedness can save the life of your patient.

Takeaway

When you come across an injured person in the wilderness, survey the scene of the injury so that you do not cause more harm or get injured yourself when helping the patient. Assess the surrounding area for any persisting dangers. Call EMS if the patient is severely injured.

Meanwhile, put on your PPE, obtain consent, and do a quick assessment for spinal injury. Then, look for whether the patient is breathing or not, or has a pulse or not (see pg. 40, 51). If the patient is not breathing, is breathing abnormally due to respiratory problems, and/or has no pulse, start CPR immediately.

If the patient is stabilized and becomes conscious, move on to head-to-toe examination. Avoid missing any injuries. Take blood pressure, temperature, and respiratory rate (vital signs) at frequent, regular intervals. If any injury is believed to be caused by long-term illness, then start with ABCDE, followed by SAMPLE instead of head-to-toe examination, and then move on to checking all vital signs. All these steps sum up your initial assessment.

But most of all, stay calm during the whole process. You may find it distressing if this is your first experience, but stay prepared and confident. Remember that you are the only help your patient has at that moment.

4

POTENTIAL BODY INJURIES AND TREATMENT

Now that you know what assessments are to be done, we are going to walk you through step-by-step potential injuries from head to toe that you or your patient could be showing symptoms of and how to treat them when we are left with our basic first-aid kit and surrounding environment.

The sequence of injuries and their subsequent management will be discussed, starting with the head, and we will move further down to other individual body parts.

Head Problems

If a person encounters a head injury, immediate danger signs need to be assessed. These signs are:

- Unconsciousness for more than 2 minutes
- Severe headache
- Loss of coordination or unclear speech
- Persistent nausea and vomiting
- Bluish discoloration behind the ears or around the eyes (signs of skull fracture)
- Blurring of vision
- Draining of clear fluid from nose and/or ears
- Seizures
- Relapse (reversion back) into unconsciousness (Centers for Disease Control and Prevention n.d.)

If these danger signs are present, urgent surgical intervention is required.

In cases of head injuries, the patient may initially seem well. However, within 72 hours, he or she may suddenly deteriorate; this time period is known as the lucid interval (Reilly et al. 1975).

If the patient improves in 48 hours, the prognosis is very good. Patients with moderate to severe head injuries (injuries accompanying the danger signs) should ideally be hospitalized, as then they can be given antibiotics, steroids to reduce brain swelling, and medicine to reduce the pressure of fluid in the brain. If the symptoms increase, they may require surgical intervention.

If you suspect spinal injury in any person with a head injury, apply a cervical collar or splints (see pg. 34, 35) to restrict neck movement. Doing so can prevent the patient from becoming paralyzed.

Vomiting in an unconscious person with a head injury can cause complications in the lungs if the particles enter the airway. To avoid these issues, place the patient face down and turned to one side, or sit the patient up at an angle of 30 degrees.

For mild head injuries, care can occur at home. Overall, the patient must rest and relax. In addition, you can:

- Ice or cool the area that was specifically injured
- Watch the patient closely for any changes, especially within the first 24 hours
- Make sure they avoid alcohol

- Give the patient painkillers for a headache if they do not make them sleepy
- If the patient is a child and they go to sleep, wake them every 4 hours to see if their symptoms have worsened
- Ensure the patient does not resume daily activity level until feeling better
- Make sure they avoid food and drink for the first 6 to 12 hours

    o   Give small amounts of food and drink when the patient starts to eat again

Know that the following symptoms are normal after a mild head injury:

- Longer reaction time
- Memory loss regarding how the injury itself occurred
- Feeling more tired than normal
- Headaches, dizziness, and thought problems
- Mood changes and difficulty focusing and completing complex procedures

Many people recover in a few days. If these symptoms continue, once available, see a doctor (Betterhealth.vic.gov.au 2014).

Neck Problems

The neck is an important part of trauma management. The back of the neck contains the spinal cord (the main nerve of our body) and vertebrae (backbone). The anterior or front of the neck consists of the airway and major blood vessels (tubular structures carrying blood to the brain). Hence, in all head injuries, these major aspects of the neck have to be looked at, too.

Anterior Neck

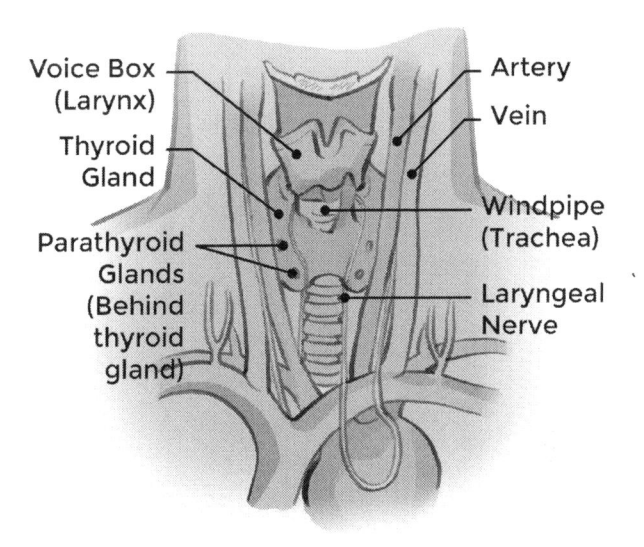

- Voice Box (Larynx)
- Thyroid Gland
- Parathyroid Glands (Behind thyroid gland)
- Artery
- Vein
- Windpipe (Trachea)
- Laryngeal Nerve

Look for point tenderness of the cervical spine (pain from touching the back of the neck). If this is present, a bone injury is suspected. In some cases, the fractured fragments may press on the spinal cord, resulting in numbness and paralysis (inability to move a part of the body). A fractured vertebra may take up to 8 weeks to heal.

Without an X-ray to confirm it, the only other option is to immobilize the patient at the suspected site of injury. If the patient is conscious (see pg. 41), for the neck region, the best option to prevent movement is to apply the cervical collar. However, if that is not available, use pads or firmly hold the head with your hands and ask the patient not to move (see pg. 34). If the neck is at an abnormal angle, gently pull on the head with your hands and align the neck in its natural position (see illustration pg. 36). You can then apply the cervical collar or tuck pads on either side of the neck to prevent its movement at rest (see illustrations and text on pgs. 34-36). Keep them in this position until help arrives if possible. This helps in protecting the spinal cord so that you do not paralyze the patient (Healthywa. wa.gov.au n.d.; Clinical Quality & Patient Safety Unit 2019).

If the patient is unconscious (see pg. 41) and a neck injury is suspected, if needed, after stabilizing them in terms of ABCDE while doing your best not to move their neck, place them in the recovery position (see pg. 36) and try to keep them there until help arrives (Nhsinform.scot 2022; Bennett 2019).

The patient can be given pain medications for pain control. In a hospital setting, they would also be given a dose of steroids to decrease the swelling around the spinal cord.

The anterior (front) neck is also to be properly inspected by looking at it. It is important to determine the depth of cuts and lacerations to learn information about the status of the voice box and the airway. The front of the neck contains blood vessels that transport blood to the brain. If injury to these occurs, it can result in severe damage to the brain and death in a very short time. Proper exploration of the anterior neck should be

Potential Body Injuries and Treatment

completed to check for injury to the voice box, food pipe (behind the trachea), and blood vessels (labeled artery and vein) (see illustration on pg. 79 for anatomy of the neck). If injured, pack the wound with gauze and ointment. Take measures to control bleeding (see pg. 169). If you don't feel a pulse, start CPR immediately.

Focus on keeping the patient's vitals stable via CPR until professional help arrives, or even if help is not available (Cheung and Napolitano 2014).

Eye Problems

Needless to say, the eye is an important organ, as it provides sight. Hence, vision must be preserved in case of its injury. Symptoms like pain, irritation, double vision, swelling, and bluish discoloration around the eye are dangerous.

If there is infection in the eyes due to germs like bacteria or viruses, the eye may have redness, be watering, and in pain. Take measures to decrease light exposure: ask the patient to wear dark glasses or a wide-brimmed hat. You can also close the eyelids of the affected eye with simple strips of tape placed vertically. Do not patch the eye, as it confines the infection, allowing it to worsen.

Wash the eye with clean water and dab with a wet, clean cloth every 2 hours. This removes pus and excess secretion. Also apply antibiotic drops and ointment until the redness, pain, and discharge from the eye improve. For an alternative eye relief and ointment option, see pg. 84.

Serious eye injuries that risk loss of vision, like those occurring from a sports injury or accident, require padding of both the eyes. The movement of one eye will decrease if only one eye is patched. Eye padding should be removed or changed at least every 24 hours. A severe blow to the eye may cause temporary blindness in both eyes, which resolves in a few hours to days.

Padding of Both Eyes

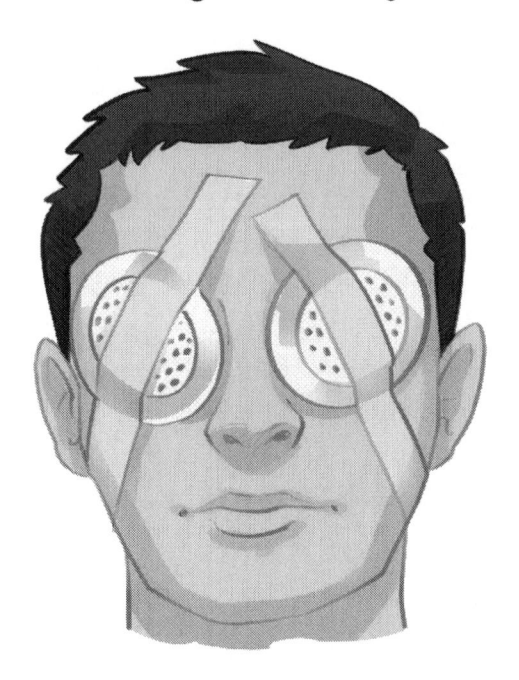

Potential Body Injuries and Treatment

If a foreign body has been removed from the eye or the outer layer of the eye is damaged, pressure patching of the eye is considered to be an effective option. For pressure patching, place two gauze pads over the affected eye. Three pieces of one-inch-wide tape are used to hold the gauze in place. The first piece of tape is fastened from the center of the forehead, diagonally downward to just below the cheekbone. The second and third strips are applied parallel to the first strip, to the left and right (see next illustration). This will provide rest to the eye and also has a somewhat protective function.

Pressure Patch

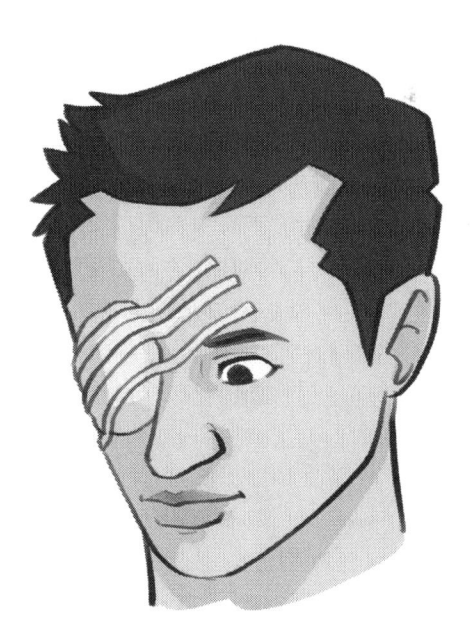

However, the comparison of pressure patching vs. antibiotic ointment has shown mixed results. So, if pressure patching is not possible, you can go for both antibiotic drops and ointments to treat cases of foreign body removal (one drop in the affected eyes 4 times a day for 7-10 days until symptoms resolve, and once before going to sleep for 7-10 days, respectively). Ointments are effective but cannot be applied during the day as they cause blurring of vision (Le Sage, Verreault, and Rochette 2001).

Note that if you do not have traditional ointment, drops of tea squeezed from a cool, non-herbal tea bag have proven to soothe an irritated eye and offer pain relief (Heimark 2010).

Ear Problems

When trauma (injury) to the ear occurs, it may either result in cuts, infection of the ear canal, or infection of the eardrum. Cuts to the ear, if involving the cartilage (a strong, flexible substance occupying almost the entire ear), can be dangerous in terms of infection. If infection progresses, it can lead to the death of the cartilage, also known as cauliflower ear. This can be prevented by cleaning the wound using sterile measures (antiseptics and sterile gauze) and bandaging, and if possible, suturing (sewing the wound to bring together the edges) is to be done within 6 hours.

The other most common causes of ear pain are infection, allergies, a foreign body, dental pain, or pain in structures of the mouth like tonsils.

Infection of the ear canal is confirmed if pulling on the ear results in excruciating pain. You can use ointment by applying it to cotton and inserting it into the canal (the big hole in the middle of your ear). Oral antibiotics can also be taken, if available.

For infection of the eardrum (the membrane in the ear which vibrates to produce sound), oral antibiotics are the mainstay of conventional treatment. Suspect that you have an infection of the eardrum if you have ear pain with or without ear discharge. If you have ear discharge, you can use an ear drop containing both antibiotics and steroids (3 drops 3 times a day). If there is only pain with no discharge, ear drops have no role. Pain medicines can be taken to relieve pain. Use these medications until the pain and discharge are gone.

Warm and cold compresses are also used to decrease earache. Alternate between warm and cold every 10 minutes until symptoms subside.

Apart from medicines, some natural remedies have also been shown to decrease ear pain. These are ginger, garlic, and naturopathic ear drops. Ginger relieves inflammation, thus reducing earache. Apply ginger juice or strained ginger oil (strain warm oil with ginger in it). Gently rub it in the ear canal (the hole in your ear) with your finger, but do not use it as an ear drop.

Garlic reduces pain and also has antibiotic properties. Soak garlic in warmed olive or sesame oil for several minutes.

Then strain the garlic out and apply the oil to the ear canal (Goldman 2018).

Nose Problems

Nasal Congestion

Nasal congestion or stuffiness may be caused by allergies or bacterial/viral infections of the nose. Bacterial infections may cause fever, yellowish discharge, and facial pain. Conventional treatment of bacterial infections involves antibiotics, but in their absence, various natural remedies are also practiced for the treatment of nasal congestion. These are saline irrigations, fluids, a humidifier, steam, warm compresses, a Neti pot, herbs and spices, and essential oils.

Saline irrigations can be prepared by adding one tablespoon of salt and baking powder each to one liter of clean water. This liquid is then sprayed into the nose with a syringe. It helps to thin the mucus in the nose and also helps clean the nasal cavities (the space inside the nostrils). A Neti pot also uses a similar mechanism. In this, a teapot-like container is used which contains saltwater and baking soda. The spout is inserted into the nostril and the cavity is flushed. The solution will then either flow out from the other nostril or out of the mouth. The flow of the Neti pot is somewhat similar to that of the syringe. However, the syringe should be refilled a number of times, which is not the case of a Neti pot. Both the syringe and

Neti pot flush can be done 3 times a day for 2 weeks until the patient feels relief from nasal congestion.

Humidifiers, steam, and warm compresses all work using the same mechanism. Heat reduces inflammation and helps unclog a stuffy nose. However, care should be taken to avoid burn injuries.

Spicy food helps the body make more mucus. Thus, it leads to a runny nose, which in turn can help cure nasal congestion. This method is only temporary and there is no available recommendation to be done on a daily basis. But we surely do not recommend it often, because we do not want you to develop stomach issues!

Essential oils like eucalyptus and rosemary, used as a nasal spray, have also been shown to reduce the symptoms of nasal congestion. How much of these should be used and for how long is still being researched (Lockett 2019).

Viral and allergic causes of nasal congestion can also be treated by supportive treatment like rest, fluids, anti-fever medicines (acetaminophen or paracetamol), anti-allergens, and pain medicines. Most of these patients improve in 7-10 days (Fokkens et al. 2020).

Foreign Body

Foreign bodies can either be inserted accidentally or intentionally out of curiosity. Children mainly present with beads, grains, or peas in their nostrils. If a child displays discharge from only one nostril, a foreign body should be suspected. In relation to the wilderness, adults can present with a leech (an insect that sucks blood by adhering to the human body; see next illustration) in the nose when drinking water from a flowing stream.

Leech

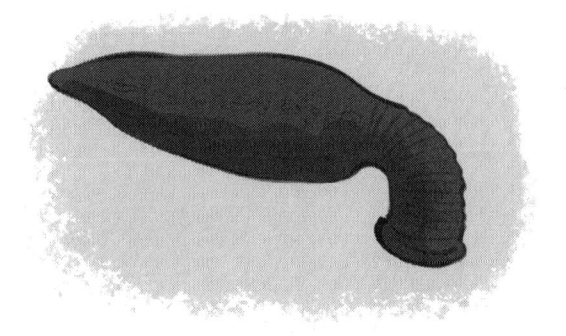

All foreign bodies should be removed, as they can lead to infection. Ideally, a nasal speculum (an instrument used to visualize the nasal cavities) is used to remove a foreign body. However, if we do not have one, we can use the end of pliers or other instruments to separate the nostril. Then, shine a light and try to discern the foreign body. If it's visible, remove it with forceps (an instrument with two prongs for grasping objects). Alternatively, flushing of the nasal cavities (nasal irrigation) can be done to wash away the debris. See pg. 86 for the details on nasal irrigation.

Do not push the foreign body backward into the mouth of children, as it may cause choking.

Fracture

The nose is one of the most common structures of the face to be fractured due to its large size. Trauma to the nose results in swelling, fracture, and nasal bleeding. Most nasal fractures are single-line, not displaced (moved from their natural position), fractures. In this case, we can go for conservative management with pain medications and a cold compress. For cold compression, you can apply ice cubes wrapped in a clean cloth for approximately 10 to 15 minutes at least 4 times a day until the swelling subsides. However, if you hear a crackling sound when pressing on the nose during the examination, try manipulating it with your hand to straighten it. Do not use excessive force. All fracture cases should undergo reduction, i.e., manipulation, of the fractured segments to the normal position (if necessary) only after the injury has been present for a week. This is the time required for swelling to decrease so that we can evaluate if the nose is deformed.

Avoid packing the nose with gauze or cotton if the nose is bleeding following trauma, as most bleeds resolve easily. Instead, pinch the nose for several minutes to stop the bleeding.

Nosebleeds

Nosebleed is a manifestation of trauma (physical injury), infection, or bleeding disorders. The initial maneuver done to

89

stop nasal bleeding is also known as the 'Trotter's maneuver," in which the person pinches their nose and bends their head forward with the mouth open. This allows them to spit out the blood so that we have an estimate of the amount.

Some other natural alternative remedies include: soaking a cotton ball in apple cider vinegar for 5 to 10 minutes and then placing it in the nose; applying ice packs or cold compresses to the nose; eating vitamin K or green leafy vegetables; putting Vaseline in the nose; increasing vitamin C by eating more foods like apples, garlic, and watermelons; and drinking enough water to stay hydrated (Rana 2018).

We should always advise the patient not to eat or drink hot food or fluids for a few days following the nosebleed. Hot food or fluids can expand the bleeding vessel, resulting in more bleeding. Also, never blow your nose. This increases the pressure and may cause re-bleeding from the blood vessel (Kucik and Clenney 2005).

Heart Beating Too Fast: Tachycardia

A rapid heart rate may signify impending shock (decrease in blood pressure with an increase in heart rate). Hence, the underlying cause of shock should be treated. This may include fluid replacements if excessive blood has been lost or the patient is dehydrated.

Other causes of rapid heart rate may be fluids in the lungs or irregular heartbeat or rhythm. Patients with fluids in the lungs

Potential Body Injuries and Treatment

experience difficulty breathing and those with heartbeat irregularities may present with rapid beating or a "funny feeling" in the heart.

If fluids fill up the lungs, the patient has to be in a propped-up position (i.e., seated) for ease of breathing. Pain medications should be given.

If someone is experiencing a rapid heartbeat, their pulse can go up to 140 to 220 beats per minute. To counter this, we can excite the vagus nerve, which is the nerve that slows down your heart.

Some of the vagal maneuvers to slow down your heart are:

- Carotid massage. Your carotid artery arises from the heart as a single artery that divides into two branches in the neck. This splitting is at the neck just below your jawline. Feel for the carotid pulse (see pg. 52 for illustration) and gently press on the enlarged portion of this artery. Always remember to feel for the pulse and massage the artery for 15 seconds, one side at a time.
- Holding your breath and bearing down very hard for 15 seconds. Bearing down refers to the use of abdominal muscles for breathing with the nose and mouth closed, as that of going to the bathroom and trying to push out while holding your breath.
- Closing the eyes and pressing firmly on one of the eyeballs for 30 seconds.
- Inducing vomiting by inserting fingers down the throat.

- Taking a deep breath and plunging one's face into ice water for 30 seconds (Arnold 1999).

All of these maneuvers decrease the heart rate. These maneuvers are to be done at the time when your heart beats very quickly as a temporary measure. However, make sure to consult your physician as soon as possible for the proper diagnosis and medical treatment.

Heart Beating Too Slow

When our body attempts to maintain blood pressure, as in the case of shock, it increases the heart rate so that blood pumps out frequently from the heart. However, there is another safety mechanism that prevents blood pressure from going too high. These sensors relax the blood vessels throughout the body and lower the heart rate. This slow heart rate can sometimes result in fainting (see pg. 153).

The heart rate can also decrease as the body temperature decreases. However, in diseases like typhoid (see pg. 116), the pulse rate is slower than would be expected for the elevated body temperature caused by the disease (Forgey 2020).

Chest Trauma

Chest trauma or broken ribs can occur due to a direct blow to the chest. Patients feel pain at the lightest touch and also on deep breath inhalation. If a rib fracture is suspected, we can simply tie a towel or large band around the chest at the site of fracture. This prevents movement and allows the fracture to heal. Generally, a fractured rib may take 6 to 8 weeks to heal. A tear in the muscles between the ribs can heal relatively quicker, i.e., in 3 to 5 weeks. Rest of the fractured segment is the most effective treatment for rib fracture.

If several of the ribs are broken in multiple places such that a segment or portion of the rib cage is detached, this segment shows paradoxical movement with respiration. This means that when the chest should expand during inhalation, this segment moves in, and when it should contract during exhalation, it moves outward. This motion is known as a "flail chest." Treatment includes placing a rolled cloth against the flailed portion to stabilize it. This is then secured in place.

Treat all of these cases with pain medications and avoid movement as much as possible. Use ice packs for pain relief if necessary. Allow the patient to sit up and place a hand over the chest when taking a deep breath. This is comfortable for the patient.

As much as we want to prevent movement of the rib cage, we also want the lungs to stay healthy. Patients often tend to breathe shallowly, as there is pain on deep inhalation associated with the fracture. This in turn can lead to pneumonia due to reduced lung function. Hence, deep breathing exercises are important to practice. Ask the patient to take a deep breath and hold it for 10 seconds, then exhale slowly. Repeat this cycle 5 times in a single sitting. This process prevents chest infection and can be repeated every 2 hours. The patient can perform these exercises until he or she is able to take a deep breath without feeling pain. A device known as a spirometer also allows deep breathing in and out to assess and improve lung function. The patient should take deep breaths 10 to 15 times every 1 to 2 hours until they stop feeling any pain (Roland 2019).

Abdominal Pain

There are various causes of abdominal pain, and this symptom can be difficult to diagnose, even for medical professionals. The most common cause of abdominal pain is gastritis, which is irritation caused by the gastric juices in the stomach. The person complains of pain or a burning sensation in the upper part of the abdomen, on the left side, or on the midline. It may be accompanied by nausea and vomiting.

If allowed to progress without proper treatment, gastritis can progress into an ulcer. A stomach ulcer is the wearing away of the lining of the stomach due to constant inflammation. An ulcer can also develop in the duodenum, which is the upper part of the small intestine. Stomach ulcers cause pain in the left side or the midline. However, a duodenal ulcer can cause pain in the right upper abdomen. It has been shown that some relief is obtained when we press the hand in the area of the ulcer.

Another possible cause of pain in the abdomen is pancreatitis, which is inflammation of the pancreas, the insulin-producing organ. Insulin regulates the blood glucose (sugar) levels in our bodies, and the body operates best when there is an optimum level of sugar. Further potential causes are gall bladder stones (in the organ producing bile, which helps in the digestion of fatty food, which in turn gives us energy), or inflammation (a response of your immune system to infections or foreign objects).

The mainstay of treatment for gastritis and ulcers is antacid medication, which should be taken until the symptoms are alleviated. The dose of the medicine and the frequency depend on the medicine used. Various medications for gastric pain relief are available in pharmacy stores as over-the-counter medications.

If there is prolonged vomiting or blood in stool that accompanies abdominal pain, try medications or natural remedies to address pain and vomiting and seek professional help.

There are various natural remedies that can be tried for abdominal pain. These are specifically helpful in the early stages and when no professional help is available.

These are:

- Ginger: Ginger is seen to have an anti-inflammatory effect and to be an effective treatment for nausea. It can be taken either in supplement form or as a beverage.
- Chamomile tea: Chamomile is also an anti-inflammatory and acts to relieve pain and cramping.
- Peppermint: Peppermint has menthol (a compound present in mint which gives it a strong smell) in its leaves, which is a natural pain reliever. It also helps fix nausea and upset stomachs.

- Apple cider vinegar: The acid in apple cider vinegar helps decrease starch digestion, allowing the starch to reach the intestine. It keeps the bacteria in the gut healthy.
- Heating pad: Heat from a heating pad, electric blanket, or hot water bottle relieves muscle spasms and improves pain. A spasm is the sudden squeezing of your abdominal muscles, which can be a symptom of abdominal pain.
- BRAT diet: BRAT stands for Banana, Rice, Applesauce, and Toast. This bland combination is great when you have to eat something in spite of nausea. It helps with nausea (the sickening feeling that vomiting will soon occur) and diarrhea (Schofield 2017).

Various studies have been conducted using varying amounts of these substances. However, there is no guideline as to the exact amount that should be taken in a day. You can try taking one or two servings per day of any of the first four in the list as a beverage. Take these until you feel better.

Apart from these, lifestyle changes are also required. Avoid alcoholic drinks, spicy foods, and acidic foods like tomatoes. All of these increase your stomach irritation and have a negative effect.

Pelvic Pain

The pelvis, also known as the hipbone, can be the site of injury during trauma and road traffic accidents. Patients can either suffer a fracture of the femur (thigh bone between the knee and the hip—the strongest bone in our body) or an anterior (front) or posterior (back) dislocation of the hip joint. Dislocation refers to the condition when the joint shifts from its normal position permanently unless fixed.

If a femur fracture occurs, the patient will have swelling and bruising in the thigh, difficulty moving the leg, and an inability to stand or walk. There is a high risk of internal bleeding and a very high risk of death. Ideally, all cases of bone and joint problems should be diagnosed with an X-ray. However, as we are talking about scenarios in this book in which professional help is not available, an X-ray is not possible. In this case, start treating a patient with a fracture of the femur with the same treatment as for shock. An adequate amount of fluids (i.e., water) is to be given if the patient can take fluids orally. Movement of the part should be prevented by traction (see next illustration). Pain medications are to be given in adequate amounts.

In-line Traction for Femur Fracture

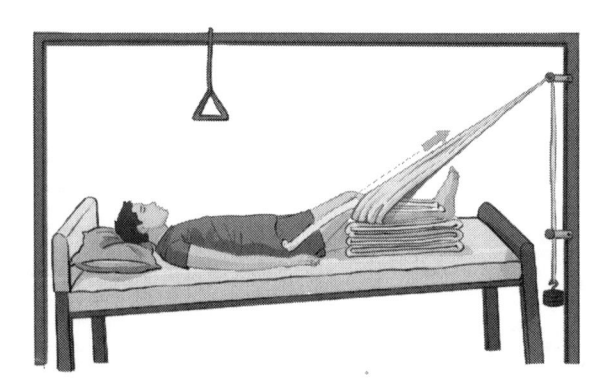

The femur is surrounded by strong muscles. In the event of a fracture, these muscles can spasm, which can displace the fractured fragments. Hence, gentle in-line traction (pulling the injured limb slowly and gently until in normal body alignment) is required. For this, the patient can be kept on a padded wooden log or a comfortable bed with a rigged footboard. This helps in avoiding skin infections and possible bed sores. A cloth is tied around the calf with an appropriate weight of about 1.6 kilograms (you can use rocks or bricks approximately this weight if you are outdoors). It takes around 8 to 12 weeks for a fractured segment to firmly stabilize (Forgey 2020).

Other causes of pelvic pain are posterior (back), central (middle), and anterior (front/forward) dislocations of the hip. Hip dislocations can also present with hip joint swelling, inability to bear weight on the leg, inability to move the legs, and

Potential Body Injuries and Treatment

decreased sensation (sensitivity to touch or temperature) in the legs. Posterior dislocation of the hip can injure the sciatic nerve, which is the most important nerve in the leg. This can cause pain in your lower limbs along with reduced sensation. Reduction (a procedure to realign a bone back to its natural position) should be undertaken within 24 hours. Pain medication can be given, if available. In order to bring the joint into the natural position, place the victim on his/her back with the knee and hip at a 90-degree angle. The femur (the longest bone that makes up your thigh) should point vertically upward. The thigh should be pulled steadily upward while simultaneously rotating the femur outward. Ideally, to perform this maneuver, two people are required (one to stabilize the body and another to manipulate the leg) and reduction may be less effective if performed alone.

Reduction Maneuvers for Hip Dislocation

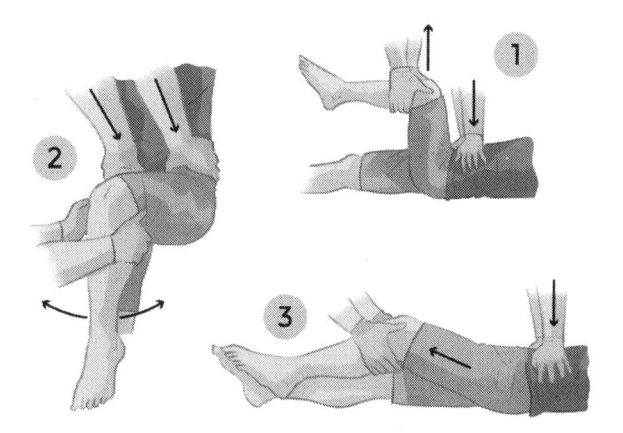

A central fracture or dislocation can present with pain, swelling, and shortening of the injured leg. It can be left as it is when medical help is not available, as the mainstay of treatment is a diagnosis by X-ray and surgery. You can splint the thigh and lower leg (see pg. 106) to prevent movement. After 3 weeks, movement and walking with crutches can be encouraged. Once they feel better, the patient can start moving on their own.

Anterior dislocation results from forceful injuries to the leg. The leg is tilted outward with pain and swelling in the hip joint. The treatment for this is reduction. The required reduction is similar to that described for posterior dislocation, but here the femur is rotated inwards rather than outwards. See the previous illustration.

Painful Testicle

A painful testicle (sex gland located between a man's legs that is ovoid in shape) can occur either with or without swelling. If there is pain along with swelling, it may be due to infection of the testicle or infection of the sperm-collecting system. If this is suspected, antibiotics should be started, and pain management involves painkillers. An antibiotic is given for 7-10 days and the dose differs according to the antibiotic used (see pg. 268 for alternative antibiotics).

Additionally, for pain relief, the movement of the scrotum (the pouch in which the testicle resides) should be prevented. In patients who can walk, provide support to avoid movement. You can do this by making the patient lie down on their back for

2-3 hours a day and by making a sling out of cloth by tying its ends around both thighs such that the cloth wraps the scrotum like a cup. This provides support for the scrotum. Continue wearing this until the swelling decreases. Cold compression can help initially for 10-20 minutes until the cloth support is ready.

Scrotal Support

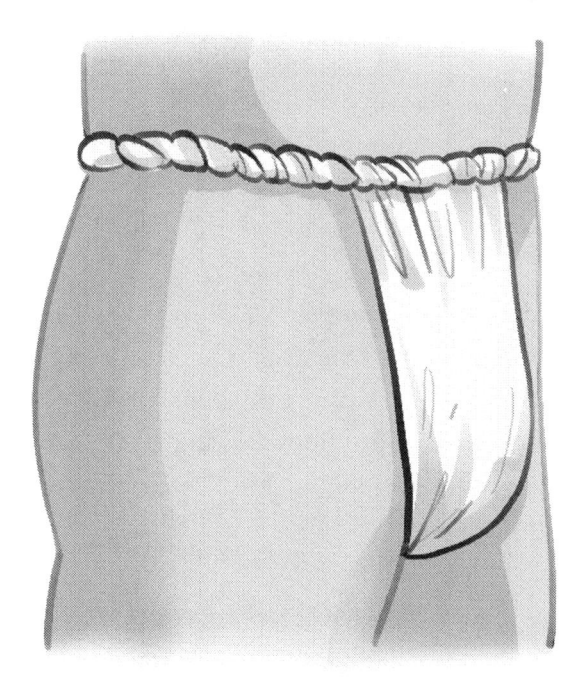

Sometimes, severe pain in the testicle may be caused by rotation of the testicle, which is known as testicular torsion. This is a surgical emergency, and hence, the patient should be evacuated as soon as possible. If timely treatment is not undertaken, it can result in sterility (loss of the reproductive ability of that organ), worsening of infection, and tissue death. This can be life threatening.

However, if evacuation to a medical facility is not possible, we can try reduction of the twisted testicle. As we know, reduction is the procedure to manipulate the body part into its own natural position. As the testicle is naturally rotated inwards, we can try to rotate it outwards, once. If pain is relieved with this simple step, it means that reduction has been achieved and further management may not be necessary. However, if pain still continues, surgical intervention is needed.

Joint Pain

Two bones meet at a joint. Joints provide a connection between bones, allow movement, and make our body flexible. The movement can either be a hinge joint, allowing movement in one direction, a pivot joint, allowing rotating or twisting motion, or a ball and socket joint, allowing the greatest freedom of movement.

If there is joint pain without any sudden trauma, it can either be due to tendon inflammation (swelling of the tissues connecting muscle to bone) or joint space inflammation. The treatment of both of these is the same. Pain occurs due to overuse or unusual compression. The general measure for relief is decreasing the use of joints and tendons or using them in a more comfortable position. If there is a sudden increase in pain, for the first two days, ice compression is of benefit (10 minutes every hour). After 2 days, heat application is beneficial and can be done for 20 minutes, 4 times per day. Creams containing painkillers alleviate the pain and provide relief. You can apply this over the painful area 3-4 times a day until the symptoms go away.

In cases of injury to the joints, injury to the bones, cartilage, or supporting muscles and ligaments may also occur. It will result in pain, swelling, bluish discoloration, and difficulty or inability to move the injured part. The best treatment for this is immediate surgical correction. However, if this fails, prevent movement of the part with splints and slings (see pgs. 106, 111) and give painkillers 3-4 times a day. By doing this, surgery can be delayed by up to 3 weeks.

The RICE method for bone fractures or joint injury relief is mentioned on pg. 71. Let us do a quick recap. RICE stands for "rest, ice, compression, and elevation." Initially, cold compression should be applied for the first 2 days, followed by heat application for 20 minutes or longer, 4 times daily (Forgey 2020). No movement should be allowed to the part and it should be kept elevated for 2-3 hours per day if possible until the swelling decreases.

Takeaway

Injuries in various parts of the body should be managed accordingly. In this book, we are assuming we do not have the facilities to conduct the required tests. It is appropriate to obtain as much information as possible about the patient's medical history along with doing a detailed examination. This can guide us towards the diagnosis. Treatment is conducted according to the availability of medicines and bandages in our first-aid kit and surrounding environment. Timely treatment will indeed help to prevent complications and ensure a good outcome.

5
USEFUL TECHNIQUES FOR HEALTH EMERGENCIES

In the previous chapters, we dealt with the physical and mental preparation needed for wilderness emergencies or a disaster-stricken household. We also dealt with the initial assessment protocol to be followed when we encounter a sick and injured patient.

This chapter explores the basic techniques of survival medicine, such as creating DIY tools and performing CPR and the Heimlich maneuver.

DIY Splints

Suppose you sustain a limb injury while on a wilderness hike or have fragments of rocks fall and injure a limb in an earthquake. This is when splints and slings best come into use.

Splints are supportive devices that protect against a sore or broken joint. They prevent bone movement, thus help with pain and promote healing. Splints can either be rigid or soft (Stang 2018).

There are ready-made, versatile splints available for purchase as first-aid equipment, or you can make them yourself. Use the illustrations on the following pages as a guide to see what your splints should look like when applying them to your leg, ankle, wrist, elbow, and thumb.

If ready-made splints are not available, you can use various improvisations. You can make splints from branches, boards, padded pack straps, or rolled-up newspapers or magazines (O'Connor 2016).

When using branches to make a splint, these are the steps to be followed:

- Place two sticks on each side of the injured hand or leg.
- Use a gauze bandage to wrap two sticks in place.
- Initially, you can tie a knot after the first wrap.
- Then, wrap the gauze bandage around the entire length of the broken segment and sticks tightly.
- A knot can also be tied in the end (Tjzawesome n.d.).

This will prevent the movement of the broken fragments.

Cushioned Do-It-Yourself Splint

Hurt Wrist

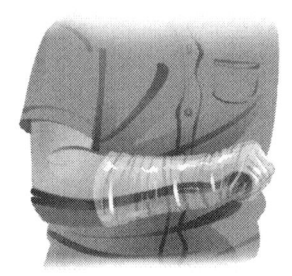

Hurt Elbow

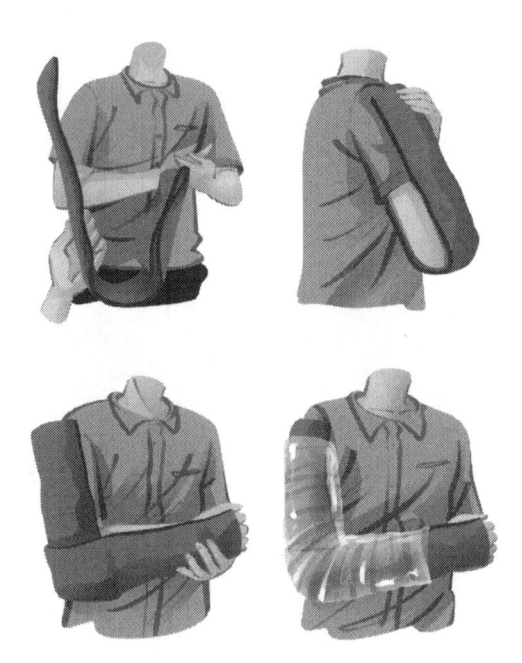

Hurt Knee: Without An Angle

Hurt Knee: With An Angle

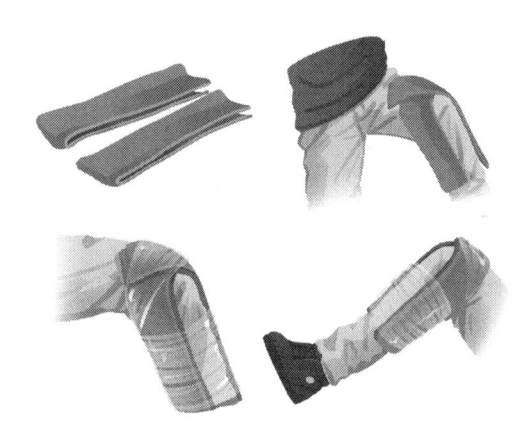

Family Survival Medicine Handbook

Hurt Ankle:
Can Put Weight On It

Hurt Thumb

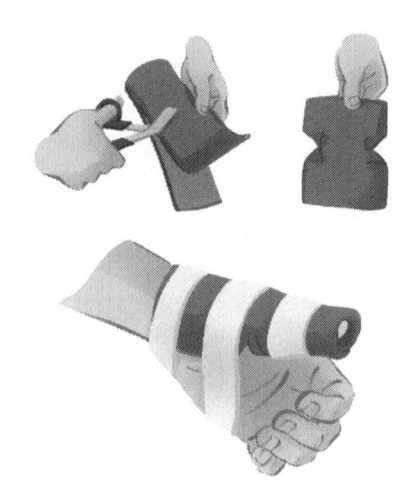

DIY Slings

If you sustain an injury to your arm or leg, there may either be a muscle injury, tendon (connects muscle to bone) injury, injured joint (see joint pain, pg. 102), or bone fracture. The first step you need to do is prevent the injured segment from moving. This will accelerate the process of the bones healing in their natural position. A sling is used for this purpose.

A sling is a flexible strap used in the form of a loop to support the weight of the injured or broken part. It connects your arm or legs to your body (Sears 2021).

A fracture is a broken bone. Your bone can break into one or multiple pieces. In some cases, there may only be a crack, whereas in other cases, some of the pieces may move from their original place and even protrude from the skin. The best way to determine if there is a fracture is by an X-ray. However, certain signs may be present which give us an idea about the presence of a fracture.

There may be swelling and pain (with or without movement) in the area of the fracture. If the fracture is in the middle of the bone, there may be bluish discoloration, indicating bleeding. The hallmark of a fracture is pain. Another way to determine the presence of a fracture is by gently twisting the injured part or compressing it lengthwise. This leads to an increase in the existing level of pain.

Now that we know how to determine if there is a fracture, let us learn how to make an arm sling, for example.

Ideally, a triangular bandage is required. But if that is not available, clothes or other bandages can also be used. The next illustration depicts how an arm sling is made (Sja.org.uk n.d.).

Do-It-Yourself Arm Sling

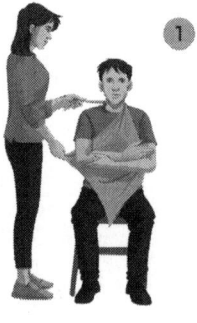

1. Make the patient sit comfortably on a chair and ask him to support the injured hand with the other hand. Take a triangular bandage or cut the cloth into a triangular shape with a 40-inch length on one side. Slide the bandage underneath the arm. One end of the triangle should lie just below the injured elbow and the next end should lie at the back of the neck of the uninjured side.

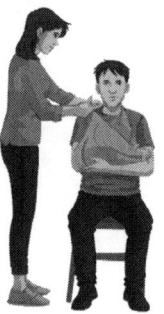

2. Hold the third end of the bandage and bring it up above the forearm. This should meet the other end at the shoulder of the injured side.

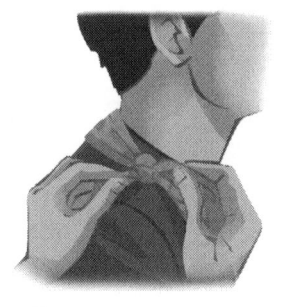

3. Tie these two ends in a reef knot and tuck in the free ends.

Useful Techniques for Health Emergencies

4. The sling should support the forearm from the elbow up to the little finger.

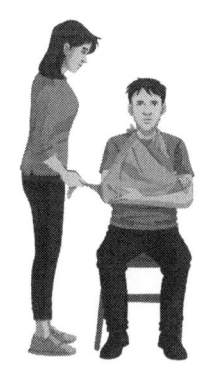

5. Fasten the end of the bandage present underneath the elbow with a safety pin or simply tuck it in.

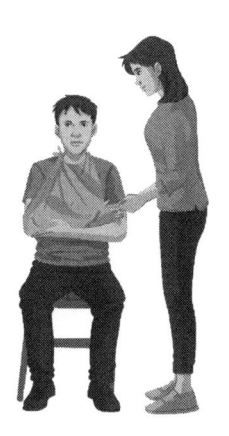

6. Check the blood flow in the fingertip every 10 minutes. You can do this by pressing the nail of the finger for about 5 seconds until it turns white and then releasing it. The reddish color of the nail should come back within 2 seconds. If not, loosen the sling.

113

Hygiene, Sterilizations, Wounds, and Equipment

Sterilization is the process to kill germs and also spores. Spores are the reproductive form of micro-organisms that lead to more germs in the future. Sterilization does not need modern facilities and can be done at home as well.

There are several ways to sterilize instruments at home:

- Initially, proper handwashing after using the restroom, before handling food, and after sneezing or coughing, use of gloves (when handling patients or any contaminated objects) and masks (when coming in close contact with sick people), timely vaccination, proper water treatment, proper food storage, and keeping insects away from households are the most effective measures to prevent disease occurrence. It is very likely for typhoid fever (see the next page for details on typhoid fever) and cholera (diarrhea due to infection) to occur in a disaster-stricken household and the wilderness. Hence, adopting the measures stated above helps prevent these diseases from occurring.
- Some instruments melt on direct heat application. Hence, for these instruments, boiling is the best means of sterilization. Materials like bandages, gauze, and tourniquets are to be boiled for at least 20 minutes before they are taken out and ready to be used. Bandages and gauze are made out of woven cotton and tourniquets are made of rubber or silicon. As these can be

damaged by direct heating, these are to be boiled. The same can be done with a T-shirt and branches before they can be used as a sling. Needless to say, if there is visible dirt on the instruments, they are to be washed off with soap and water before boiling. We can also use alcohol as a disinfectant. The item should be soaked in 60-80% alcohol for at least 30 seconds.
- Another direct means of sterilization is the use of flame to directly heat the instruments. This can only be done for materials that do not melt on heat application, such as metal forceps and trays. Hold the instrument with prongs over direct heat. Don't use metal prongs without protecting your hand, as you can get burnt. Use an insulating material such as a cloth to hold the prongs. Heat until the material turns red. Then, it can be cooled down (Nicholson 2015).

All these steps are important for first aid as well as survival medicine. Whenever any injury cuts open your skin, it forms a roadway through which infection can travel into your body. Moreover, the instruments, gauze, and padding which you use can also act as a source of infection. Hence, make sure you follow the preventive measures and use sterile dressings and instruments so that you make the wound sterile (free from germs). This will assist in faster healing of the wound and prevent the spread of germs into your bloodstream (Royallifesavingwa.com.au n.d.).

Special Note: Typhoid

Typhoid fever is caused by an organism named Salmonella typhi. This is a type of germ known as bacteria. It is transmitted through the intake of food contaminated with the bacteria. The patient presents with fever, body ache, headache, disturbed toilet habits, loss of appetite, and weakness. On taking vitals, the patient has a high fever. However, the heart rate is on the lower side in contrast to the fact that the pulse rate rises with the rising body temperature. The patient is asked to maintain hygiene and is treated with antibiotics.

Necessary Techniques for Emergency Situations

CPR (Cardio-Pulmonary Resuscitation)

CPR is a life-saving procedure comprised of chest compression and breathing assistance, administered when a patient is not breathing, is abnormally breathing, or if we suspect his or her heart has stopped. If you are not confident enough to provide breathing assistance, you can administer only chest compressions.

Administering CPR on a severely injured person does not mean he or she is sure to live. In most instances, CPR can only revive a person temporarily and it is necessary for the person to be kept on a ventilator (a machine providing breathing support in patients whose breathing has stopped), or he or she may require a defibrillator (a machine that delivers electric shocks in patients whose heart has stopped or has abnormal rhythm). Without these, the outcome is more likely fatal (Alton 2015).

However, there are also other scenarios like anaphylaxis (severe allergic reactions) where the heart stops and CPR is necessary because it gives adrenaline, which is the medicine used to combat anaphylaxis, time to work (Alton 2015). Since we acknowledge that every life is precious, we discuss CPR on a step-by-step basis, as even one life saved is a remarkable achievement.

When Should You Give CPR?

You need to start CPR immediately if:

- Your patient is not breathing
- The patient gasps for air (due to difficulty breathing, a person takes sudden, loud gulps of air with their mouth, known as gasping) or is abnormally breathing (see pg. 45) due to respiratory problems
- You can ask the patient if he/she is choking or not. If choking, refer to pg. 131. If no response or a negative response, start CPR immediately
- Their heart has stopped, thus, you cannot find a pulse
- Your patient is unconscious or unresponsive and not breathing

All these signs can occur in the following circumstances:

- Heart attack or cardiac arrest

- Choking
- Near drowning
- Suffocation
- Poisoning
- Drug or alcohol overuse
- Smoke inhalation
- Electrocution
- Road traffic accident (Martin 2022)

In all of these emergencies, begin CPR immediately, as it could be life-saving. Chest compressions in CPR help establish circulation to the brain and other vital organs. Delaying this treatment might harm the brain or other organs, which could cause further issues in the future.

How Do You Give CPR?

Now that we have learned when to give CPR, let us take a step-by-step approach to learn how to give CPR to adults (ages 12 and up).

Step 1: Does the injury appear very serious? If your answer is yes, if available, call for emergency help. Next, put on your safety gear (PPE, or personal protective equipment, which involves a gown, gloves, mask or face covering, and eye protection, worn to prevent contracting diseases that can be easily transmitted) (Redcross.org n.d.). After that, ask for consent and quickly assess for a neck injury (see pg. 33).

Step 2: Quickly start with an assessment of the airway and

Useful Techniques for Health Emergencies

breathing as a part of ABCDE, as discussed in Chapter 3 (see pg. 40). If your patient is not responsive, open their mouth and remove any visible, loose objects that may be obstructing the airway.

Removing Foreign Body from the Airway

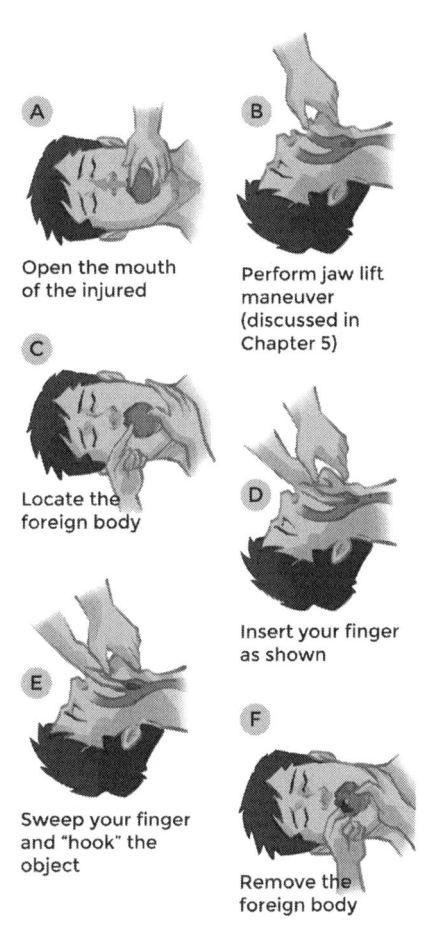

A — Open the mouth of the injured

B — Perform jaw lift maneuver (discussed in Chapter 5)

C — Locate the foreign body

D — Insert your finger as shown

E — Sweep your finger and "hook" the object

F — Remove the foreign body

Next if their airway is still not open or they are breathing abnormally, check to see if you can open the airway by using the head tilt-chin lift maneuver or jaw thrust if a neck injury is suspected.

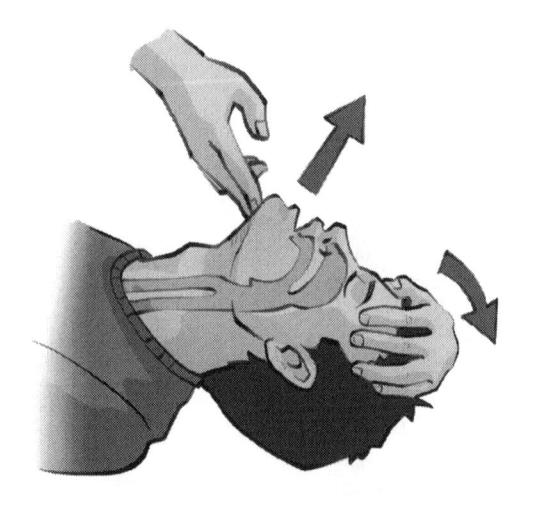

Head Tilt-Chin Lift

Jaw Thrust: If neck injury is suspected, open airway like this

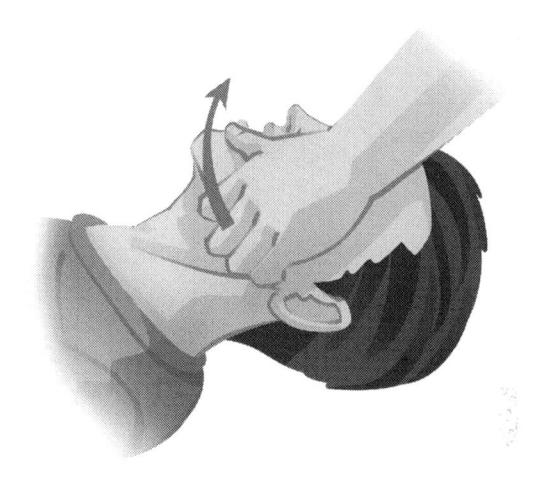

Step 3: Look for any signs that a person is breathing as attempts to clear the airway are completed. We have also discussed these in the previous chapter (see pg. 40). Assess for the rise and fall of the chest. You can also determine the movement by putting a hand above the chest. Place your fingers just beneath the nose and feel for the exit of air. If no air movement or breathing is detected, after completing step 2, start step 4 immediately (see pg. 122).

Check for signs of circulation. You can feel for a carotid (neck) pulse as seen below and as referred to on pg. 122. If there is no sign of circulation (like an absent pulse), start step 4 immediately.

Circulation?
Neck Pulse Check

Step 4: Chest compressions: 30 chest compressions are to be performed. Place one hand above the other and clasp them together. Keeping your elbows straight, push fast and hard with the heel of your hands. Hands are to be placed just below the nipples in the midline (middle of the body). Push the chest at least 2 inches deep and deliver CPR at the rate of 100 times per minute.

CPR Chest Compression Diagram

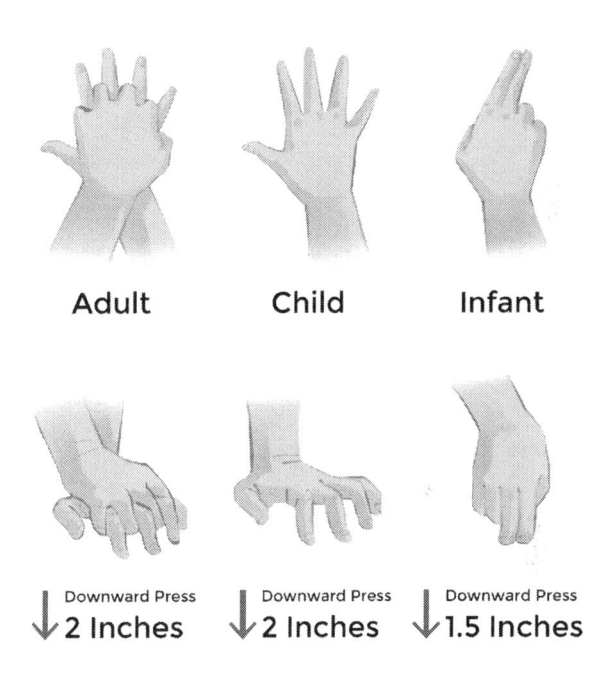

Adult	Child	Infant
Downward Press ↓ 2 Inches	Downward Press ↓ 2 Inches	Downward Press ↓ 1.5 Inches

The rhythms of certain popular songs, such as "Stayin' Alive" by the Bee Gees, can be used when adjusting the rhythm for providing CPR (see pg. 27 for other examples). Let the chest rise fully in between compressions.

Step 5: Prepare for breathing assistance: After 30 chest compressions, prepare to give 2 rescue breaths. If you are not trained in providing rescue breaths, you can opt to perform only chest compressions.

If no neck injury is suspected, the head tilt-chin lift maneuver can be undertaken, whereas if a neck injury is suspected, the jaw thrust maneuver should be used. Both of these open up the airway if there is a fallback of the tongue, which mostly occurs in the case of unconscious patients.

For the head tilt-chin lift maneuver (see the next diagram):

1. Kneel next to the head of the patient.
2. Tilt the head gently backward by placing one hand on the patient's forehead.
3. Put your index and middle fingertips beneath the bony portion of their chin, then raise their chin vertically upward. By doing so, the tongue lifts away from the airway leaving it unblocked.

Head Tilt-Chin Lift

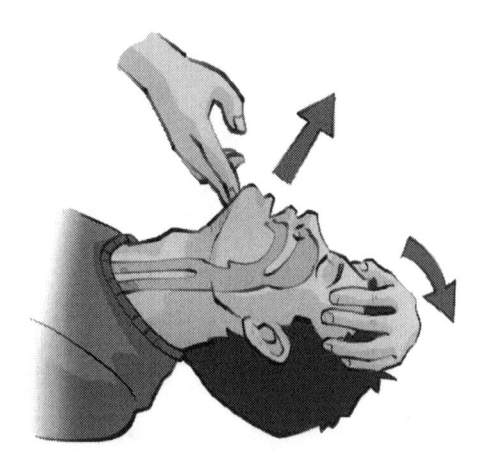

The head tilt-chin lift maneuver manipulates the neck, thus is not possible to do in patients who have sustained a neck injury. In these patients, a jaw thrust maneuver is to be done.

For the jaw thrust maneuver (see the next diagram):

Hold the angles of the lower jaw and lift it with both hands. This will move the jaw forward (Furst 2017).

**Jaw Thrust:
If neck injury is suspected, open airway like this**

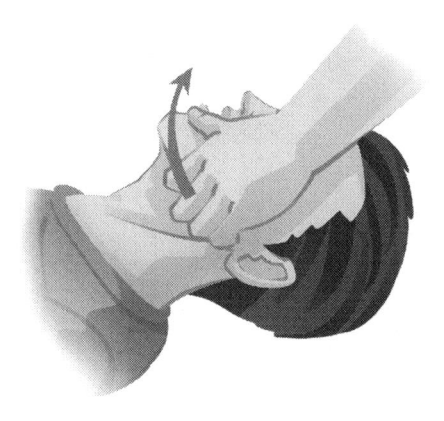

Step 6: Deliver rescue breaths while maintaining an open airway either via the head tilt-chin lift maneuver or jaw thrust technique if a neck injury is suspected. Pinch the nose of the patient tightly. If the patient's lips are closed, open them with

your thumb. Then, place your mouth fully over theirs and blow, such that their chest rises. If the chest does not rise and fall during the first rescue breath, reassess your open airway technique (head tilt-chin lift or jaw thrust) prior to giving the next one. If, after doing so, the second breath is still not received (meaning the patient's chest does not rise and fall), your patient may be choking. Clear the object accordingly before continuing with CPR.

Rescue Breathing

Step 7: Repeat these cycles of 30 chest compressions and two rescue breaths until medical help arrives or the person starts breathing (Barrell 2020).

CPR for Kids

If you find an infant or child not breathing and whose heart is stopped, start CPR immediately. Here, an infant refers to a baby less than one year old and a child refers to someone between 1 to 12 years old. The steps are somewhat similar to that of adults; however, there are significant differences.

Step 1: Does the injury appear very serious? Ask yourself, and if your answer is yes, if available, call for emergency help. Next, put on your safety gear (PPE), i.e., gloves and mask. After that, ask the parent(s) for consent, and quickly assess for neck injuries.

Step 2: Start with an assessment of the airway and breathing as a part of ABCDE, as discussed in Chapter 3 (see pg. 40). Is the child or baby conscious? If not, open the mouth of the child and remove any visible, loose foreign objects that may be obstructing the airway.

Family Survival Medicine Handbook

Removing Foreign Body from the Airway

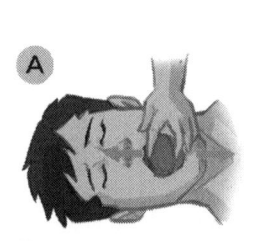

A Open the mouth of the injured

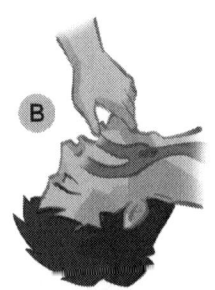

B Perform jaw lift maneuver (discussed in Chapter 5)

C Locate the foreign body

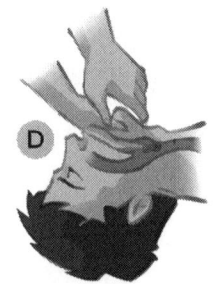

D Insert your finger as shown

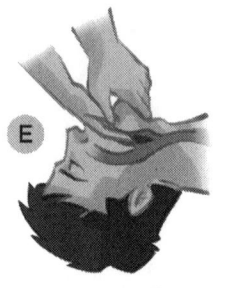

E Sweep your finger and "hook" the object

F Remove the foreign body

Useful Techniques for Health Emergencies

Check to see if you can open the airway by using the head tilt-chin lift maneuver (see pg. 124 and next illustration) or jaw thrust technique if there is a suspected neck injury (see pg. 125 and next illustration).

Head Tilt-Chin Lift

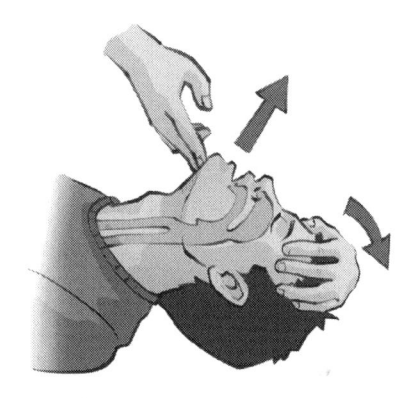

**Jaw Thrust:
If neck injury is suspected, open airway like this**

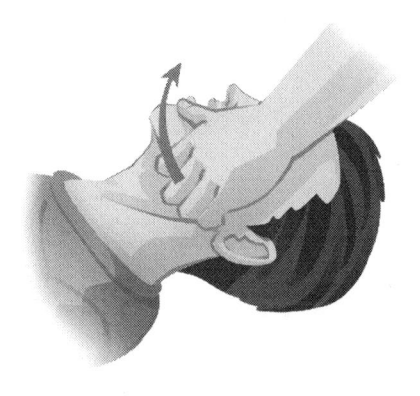

Step 3: Check for signs of breathing. These are the same as described in step 3 of adult CPR (see pg. 121). Is their chest moving? Changes in infant breathing patterns are normal as they usually have periodic (occurring in intervals) breathing. If the infant or child is not breathing after completing step 2, start step 4 immediately.

Step 4: 30 chest compressions for a child are given with the heel of one hand. In an infant, it is given with two fingers. Place the fingers or heel of your hand in the center of the chest just below the nipples. Then give compressions at the rate of 100/min, 2 inches deep for a child. For an infant, administer chest compressions at the rate of 100/min, 1.5 inches deep (see next illustration.)

CPR Chest Compression Diagram

Adult	Child	Infant
Downward Press 2 Inches	Downward Press 2 Inches	Downward Press 1.5 Inches

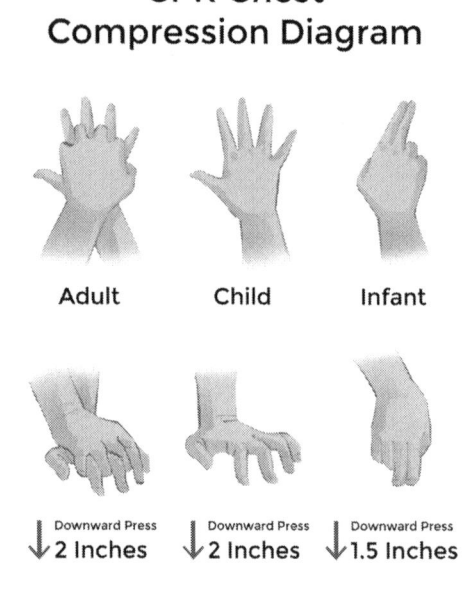

Step 5: Prepare to give two rescue breaths while simultaneously maintaining an open airway. If you are not trained in providing rescue breaths, you can opt to perform only chest compressions.

Open the airway by doing the head tilt-chin lift maneuver or jaw thrust if there is a suspected neck injury (see pg. 33). Pinch the nose of the child and blow air into their mouth for 1 second. If the child is an infant, make a seal over both the nose and the mouth and blow air into the infant's mouth. If the chest does not rise and fall during the first rescue breath, reassess your open airway technique (head tilt-chin lift or jaw thrust) prior to giving the next one. If, after doing so, the second breath is still not received (meaning the patient's chest does not rise and fall), your patient may be choking. Clear the object accordingly before continuing with CPR.

Step 6: Repeat rescue breathing and chest compression cycles until help arrives or the child starts breathing (Barrell 2020; The American Red Cross 2022; American Red Cross n.d.).

Heimlich Maneuver

The Heimlich maneuver is a life-saving procedure administered to a choking individual. Choking occurs when an object obstructs the airway, so the oxygen supply to the brain is blocked. In adults, the object is mostly food and food particles. In infants and children, it is mostly objects inserted accidentally or experimentally into the mouth. Choking with complete obstruction cuts off the oxygen supply to the brain. If there is partial airway obstruction due to choking, the patient will have distressed breathing. They cannot talk and may have noisy breathing. However, the hallmark sign is that they will have their hand clutched around their throat, indicating that they are having trouble in that area.

Other signs shown by choking individuals are:

- Squeaky sounds when trying to breathe
- Weak or forceful cough
- Pale or bluish skin, lips, and nails
- Loss of consciousness

Some patients can forcefully remove the choking objects themselves by coughing. If the patients are not in a position to expel the foreign bodies themselves, the American Red Cross recommends a "five-and-five" approach (Note: the technique changes for obese people and pregnant women. The changes are discussed later in the chapter):

- Give 5 back blows. You should stand on the side and just behind a choking adult. For a child, kneel behind. Then, place one arm across the person's chest for support. Bend the person at a right angle and deliver five back blows between the person's shoulder blades with the heel of your hand.
- Give 5 abdominal thrusts. Abdominal thrusts are also known as the Heimlich maneuver, which is further discussed on pg. 134.
- 5 back blows are given, followed by 5 abdominal

thrusts in cycles until the blockage is dislodged (Mayo Clinic Staff 2020).

Five-and-Five Method

To effectively administer abdominal thrusts, also known as the Heimlich maneuver, to a choking individual, do the following:

- Stand just behind the patient and wrap your hands around his or her waist, as shown in the illustration.
- With one hand, make a fist and place the thumb side against the patient's abdomen, in the middle, slightly above the navel.
- Grasp your fist with the other hand.
- Lift your elbows away from the victim's body and press your fist into the victim's abdomen with a separate, distinct, and quick upward thrust (Forgey 2020).

See the illustration below.

This is the standard approach for an adult patient who is not yet unconscious, i.e., we have diagnosed them as being in the early stage of choking.

Abdominal Thrusts

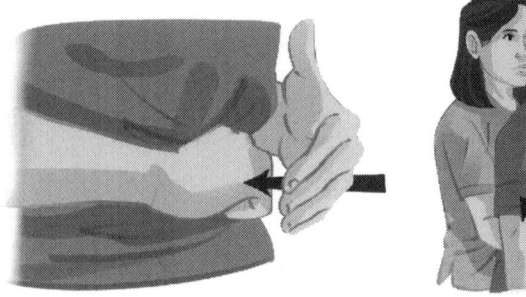

If there is complete blockage of the airway and the patient is unconscious, begin CPR (see pg. 116) immediately. In between chest compressions, look into his or her mouth for any foreign body and remove it if possible.

Removing Foreign Body from the Airway

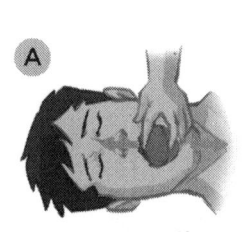

A — Open the mouth of the injured

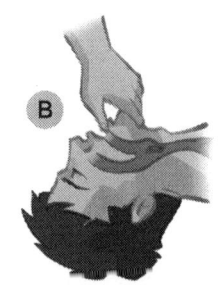

B — Perform jaw lift maneuver (discussed in Chapter 5)

C — Locate the foreign body

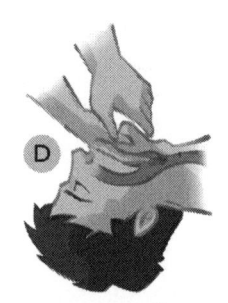

D — Insert your finger as shown

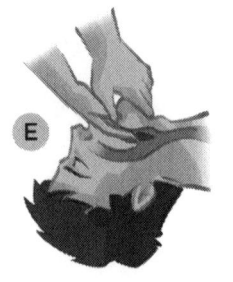

E — Sweep your finger and "hook" the object

F — Remove the foreign body

If you are alone and you choke on something:

- Call 911 immediately, if you can. Until help arrives, you have a very limited time.
- You may not be able to deliver back blows to yourself effectively, but you surely can do abdominal thrusts.
- For this, place a fist over your abdomen, slightly above the navel.
- Make a tight fist with one hand and grasp it with your other hand. Then, bend over a hard surface (a countertop or a chair).
- Shove your fist inward and upward (Mayo Clinic Staff 2020).

Self-Administered Abdominal Thrusts

For Pregnant Woman or Obese Person Choking

- Place your hand in a fist-like manner similar to that of the Heimlich maneuver (abdominal thrusts) at the base of the breastbone, just above the joining of the lower ribs.
- Then, press hard into the chest with a quick thrust, in the same manner as the Heimlich maneuver (abdominal thrusts).
- Repeat the procedure until the food or other blockage is dislodged. If the person becomes unconscious, start CPR immediately (Mayo Clinic Staff 2020).

Abdominal Thrust for Pregnant or Obese Patient

Useful Techniques for Health Emergencies

When it comes to an infant choking, telltale signs are that the infant turns bluish, has a weak cry, makes noisy breathing sounds, has a weak cough, and shows signs of forceful breathing like chest retractions (sinking in of the chest wall during breathing due to inadequate air pressure). Do not proceed with the following steps if the infant is coughing hard or crying a lot because maybe they can expel the object themselves.

The steps to be followed for a choking infant are:

- Call 911. If you have an accompanying person, call for help.
- Initially, start with the back blows. Place the infant face down along your forearm in such a way that the head is lower than the body. You can support the child in your lap. Open the jaw with your fingers. Then, use the palm surface of your free hand to give 5 quick blows in between the shoulder blades of the infant. After 5 blows, check to see if any foreign object is visible in the mouth. However, do not check by turning the infant onto their back again, as the object can again return to the airway.
- If the object is not seen in the mouth even after 5 back blows, proceed to chest thrusts. If you do see the object, try to get the baby to eject it themselves or remove it yourself only if easy to do so. In a conscious infant, do not try to grab and pull out the object. In an unconscious infant, only if the object is visible, try to remove it with a finger. Do not blindly search for the foreign object in either a conscious or an unconscious infant.

Let the infant eject the object from his or her mouth on their own.
- For chest thrusts, place the infant face up along your forearm using your thigh or lap for support, such that the head is lower than the body. Hold the back of the infant's head in your one hand. Place your two fingers in the center of the infant's chest just below the nipples. Give up to 5 quick thrusts, compressing the chest between 1/3 and 1/2 inches in each thrust. This is about 1.5 to 4 cm (0.5 to 1.5 inches). Repeat the cycle of 5 back blows followed by 5 chest thrusts until the object is dislodged or the infant becomes unconscious.
- If the child suddenly becomes unresponsive, begin infant CPR immediately (see pg. 127) (Habrat 2019).

Useful Techniques for Health Emergencies

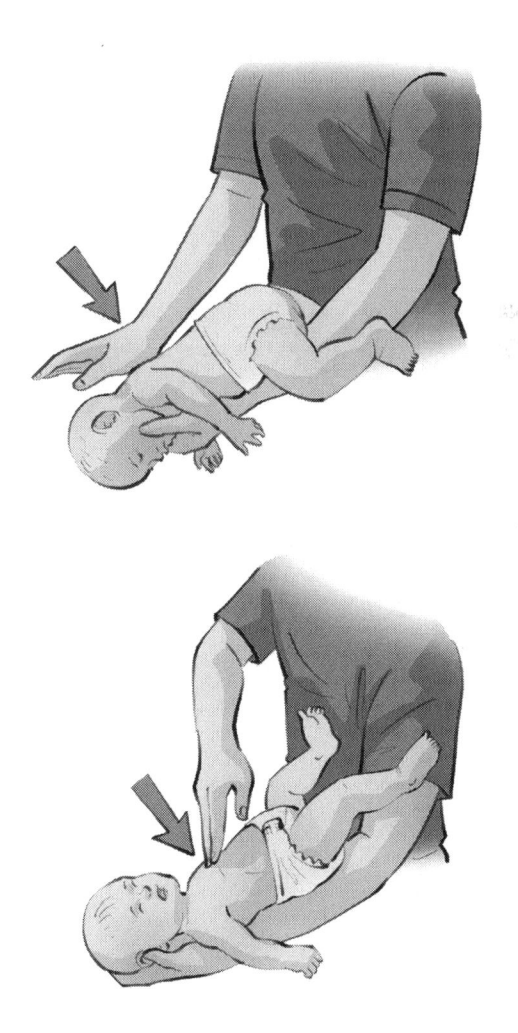

Takeaway:

This chapter gives you information about some of the most important life-saving procedures. The knowledge of these maneuvers along with the skills to perfectly execute them are necessary for an emergency scenario. This can make the difference between life and death. A single procedure may prove useful in varied scenarios. For example, CPR can be used for an unconscious non breathing person, the cause of which may be drowning, electrocution, or even a heart attack. Hence, keep learning and practicing enough so that you can use these procedures when necessary.

Help Save a Life?

We have so far learned the basic approach to assisting an injured person and some life-saving maneuvers. That said, we just wanted to ask you this: have you found any value at all in this book so far? Studies have shown how much more fulfilling life is when we help others. It doesn't matter if the person is known or unknown. Let us share two short stories with you all.

A friend of ours was dining with his family. Suddenly, his dad choked on a piece of food. He started coughing and turning blue. Thanks to the fact that our friend knew how to perform the Heimlich maneuver, he saved his dad's life. Think about how catastrophic the situation could have been within minutes.

We have another friend who was out eating brunch with her family when a one-year-old baby in the restaurant started choking. The child's mom stood up and started screaming and crying out for someone to help her because her baby was choking. Because our friend knew the Heimlich maneuver, that sweet baby lived to see another day. Let these incidents act as an eye-opener for you.

Do you want to be the person to save someone's life? Do you also want to help another family survive a medical emergency?

Consider paying it forward by sharing this book with another family. You never know who might be saved from great pain.

You could also leave a review. Maybe you also want to help

those you do not know by letting them know that they too should obtain this book to potentially save a life.

Life is short as is. So, if at all possible, please consider being the reason someone can live it out in less pain or stress, or even to live it at all.

Now, let's continue learning common survival medical techniques for when help may not be on the way.

To your continued survival preparedness.

Survival Knowledge Is Power (SKIP)

6
HANDLING MINOR HEALTH EMERGENCIES

Injuries can be divided into categories of minor, moderate, or severe. The upcoming three chapters will teach you how to handle these emergencies. This chapter focuses on low-risk emergencies.

The phrase "minor emergencies" can be misleading, as an emergency can hardly be called minor. However, in cases of minor emergencies, an emergency room is not usually required and they can be managed independently.

Minor emergencies include:

- Injuries, including sprains (injury to a ligament), strains (injury to muscle), or mild fractures
- Mild lacerations (cuts)
- Headaches
- Mouth sores
- Fainting
- Dizziness
- Prolonged cold and flu symptoms
- Stomach and small intestine upset, including nausea and vomiting
- Mild asthma attacks
- Rashes and skin lesions (change in normal skin structure due to disease) (Bellingar n.d.)

Now, let us discuss some of these in detail.

Headache and Migraine

Headaches and migraines are common occurrences. Headache, of course, refers to pain in the head. A migraine is a severe form of headache which affects the daily activities of a person and may also be associated with auras (flashes of light and color). Not all headaches occur due to a mass in the brain or an artery about to rupture; however, these should still be suspected if a patient experiences vomiting, loss of consciousness, or abnormal body movement.

If a person states that he or she has a headache, initially, maintain hydration by giving him or her adequate fluids. If the headache is severe, painkillers can be given.

A migraine is more severe than a headache and may involve light and noise sensitivity and auras (flashes of light and color). A migraine can be triggered by stress, inadequate sleep, and caffeine intake. Hence, identifying and avoiding triggers is the first step toward treatment.

For the first aid of migraine treatment, the following measures are to be adopted (Furst 2020):

- Darken the room: The patient has light sensitivity and prefers to be in the dark.
- Painkillers: These are available as over-the-counter medications, i.e., they are available without a prescription. Painkillers like paracetamol and acetaminophen

are easily available and are helpful in migraine treatment. For other alternatives, see below.
- Cold compression: Cold compression with ice over the scalp is known to relieve migraine headaches. You can wrap ice cubes in a clean cloth and use them for cold compression. Then, place ice over the scalp for 15 to 20 minutes at a time, 3-4 times a day. You can repeat until the migraine ends.
- Anti-vomiting medications: If the migraine is accompanied by nausea and vomiting, anti-vomiting medications can be taken, which are available as over-the-counter medicines.

Other causes like dental pain, jaw joint pain, and heat (see pg. 211) can also cause headaches, and these need to be individually addressed (Forgey 2020).

If dental pain is the cause, here are some at-home remedies known to alleviate pain and, if the injury is minor enough, prevent a dental visit:

- Salt water mouthwash: Simply mix a ½ teaspoon of salt (known to be a natural disinfectant) into an 8-ounce glass of warm, clean water. Swish it in your mouth and gargle for 30 seconds without swallowing.
- Hydrogen peroxide mouthwash: Mix 3% hydrogen peroxide (a bacteria-killing, inflammation- and pain-relieving household chemical) with an equal amount of clean water. Swish it in your mouth and gargle for 30

seconds without swallowing. Then spit it out. This can be done up to 4 times a day.
- Ice or cold water: Hold a clean cold towel or clean towel wrapped around a bag of ice for 20 minutes. This can be repeated every few hours. This is especially helpful if a particularly bad event caused mouth pain.
- Peppermint tea bags: Place a chilled, used tea bag on top of the painful area. This is especially helpful to cool and numb the area when gum sensitivity is involved. It should take no more than a few minutes to chill the tea bag.
- Other home remedies include chewing slowly on a fresh garlic clove or toothache plant, applying vanilla extract to the injured area, making guava leaf mouthwash via boiling guava leaves and using accordingly, and drinking wheatgrass or using it as a mouthwash (Cronkleton 2022).

Some jaw joint pain natural remedies include:

- Applying a cool cloth or ice bag or pack to the patient's jaw for 10 minutes. Rest for 10 minutes then repeat. Do this up to 4 times. This will decrease the pain and decrease swelling. Make sure it is not so cold that it damages the skin. This may require wrapping the ice bag with another clean cloth before applying it to the skin.
- Wrapping a hot water bottle with a damp towel or warm compress on the patient's jaw. Make sure not to make it so hot it burns.

- Supplements such as magnesium—which is found in almonds, spinach, avocados, chard, and bananas, to name a few—help relax the muscles; omega-3 fatty acids, which are found in walnuts, salmon, and chia seeds, have anti-inflammatory abilities; and turmeric.
- Stress management.
- Soft foods.
- Avoiding chewing gum, hard or chewy candy, caffeine, and slouching (Prescott Dentistry 2022).

Insect Bites and Stings

Insects (mosquitos, fleas, or bedbugs, for example) are known to use their mouth to break a person's skin for feeding purposes. This is known as a bite and they usually itch.

This is different from when insects sting. A sting is when an insect uses another body part such as its tail to inject something like poison (venom) into a person. They do this in self-defense.

Stings tend to hurt more than bites. However, both tend to cause short-term discomfort, be annoying, and other than that, offer no serious health problems (Kidshealth.org 2021).

It is a common occurrence to be bitten or stung by an insect in the wilderness. The symptoms you develop are determined by how poisonous the bite or sting of the insect is.

If you have been bitten by an insect, here are a few ways to alleviate discomfort and/or itching: apply ice directly on the infected area for no more than 5 minutes to avoid skin damage, or apply honey, witch hazel, rubbing alcohol, or aloe vera as needed (Fanous 2019).

If you have been stung by an insect and the stinger is still visible in your skin, try removing it by gently scraping the skin with a flat-edged object or your fingernail. Avoid using tweezers in most cases as they could squeeze the stinger and release more venom.

Stinger Removal

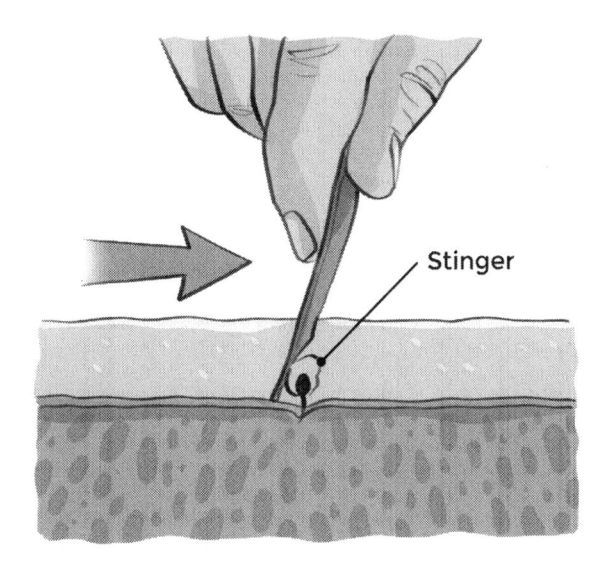

For bites and stings, wash the area with soap and clean water. Apply cold compression by wrapping ice cubes in a clean cloth and placing the compress on the affected area for about 10 minutes daily until it heals. This helps reduce pain and swelling. Apply calamine lotion to relieve itching and pain until those symptoms are gone (Hepler 2017).

If a tick gets attached to your skin, remove it by grasping it with tweezers as close to the skin as possible. Avoid the tick getting crushed or the mouth being left in the skin.

If a caterpillar's hair gets stuck to your skin, use tweezers to remove it.

Spider bites can be treated as normal insect bites. However, some species of spiders may cause an intense reaction (see below) within a few hours of being bitten. These may need medical attention (Nhs.uk 2019).

In severe cases, insect bites and stings can cause sudden swelling of the entire body followed by difficulty breathing, fast heart rate, and fainting. This severe allergic reaction is known as anaphylaxis. This requires immediate medical attention as the outcome may be fatal. A drug known as epinephrine is to be given immediately via a syringe or an EpiPen. Some people know they are severely allergic to certain bites or stings or even foods and thus carry an EpiPen with them. Make sure to ask your patient if they have this medication so you can administer it in their thigh as soon as possible. If the heart stops before the drug starts working, start CPR immediately.

Fainting

Fainting refers to the condition when a person loses consciousness and quickly becomes unresponsive due to decreased amount of blood flowing to the brain. Thus, patients may fall if they are not caught in time. Fainting may occur due to stress, exhaustion, hunger, or pain. If someone faints and does not respond for quite some time, this is serious and needs immediate medical attention. If so, call 911 and then put on your protective gear (see pg. 32). Open the airway of the patient (see pg. 40), assess if he or she is breathing (see pg. 40), and if not, start CPR immediately (see pg. 116).

If a person suddenly faints, you can:

- Ask them to lie down if they are not already doing so.
- Check for injury to other body parts where they might have hurt themselves if they fell during the fainting episode.
- Kneel in front of the patient and raise their legs above their heart by putting both their ankles on top of one of your shoulders. This process increases the blood flow to the heart. You can maintain this position until the patient regains consciousness.
- You or someone else can open windows and give fresh air to the patient after you finish the above steps.
- Reassure the patient and help them sit up slowly. Give fluids and ask them to adequately rest (Sja.org.uk n.d.).

Bringing Down Fevers

An above-normal temperature is referred to as a fever. That being said, what is a person's typical body temperature?

Specifically, each person's normal body temperature will differ, however it ranges between 97 F (36.1 C) and 99 F. (37.2 C). In some, this range may slightly be different but overall, the average body temperature is 98.6 F (37 C). The following thermometer readings indicate a fever:

- Rectal temperature: 100.4 F (38 C) or higher (There are specialized rectal thermometers for rectal temperature measurement)
- Oral temperature: 100 F (37.8 C) or higher
- Armpit temperature: 99 F (37.2 C) or higher (Mayo Clinic 2022)

We have already discussed the sites to measure temperature and how to do it. You can refer to pg. 70 for further details.

Fever itself isn't a disease but an indicator of an underlying infection in your body. The causes of a raised body temperature are:

- Infections (bacterial or viral)
- Inflammatory diseases such as rheumatoid arthritis, or inflammation of the synovium (the lining of the joints)
- Heat exhaustion

- An abnormal growth (malignant tumor) that can spread to other parts of the body
- Antibiotics, blood pressure medications, and drugs that treat seizures (sudden, abnormal body movements)
- Some vaccinations, such the diphtheria, tetanus, acellular pertussis (DTaP), or pneumococcal vaccine (Mayo Clinic Staff 2022)

The symptoms of fever in adults are:

- Chills (shivers)
- Sweating
- Headache
- Weakness
- Irritability
- Loss of appetite
- Dehydration

A child with a fever may be fussy and uncomfortable along with the above symptoms.

Prevention

Since fever can be a sign of an underlying infection, you have to prevent spreading the infection by the following steps:

- Wash your hands thoroughly before touching food items, after coughing and sneezing, and after using the restroom.

- Avoid touching your nose, eyes, and mouth at all times. If you have to, make sure you have washed your hands before and after these actions.
- Cover your nose and mouth with a handkerchief while sneezing or coughing.
- Carry hand sanitizer with you and use it frequently, as it prevents infection.
- Avoid sharing utensils, as saliva can act as a means to spread infection (Mayo Clinic Staff 2022).

Dangers

High fever can accompany several serious symptoms. These are:

- Breathing problems
- Confusion
- Abnormal body movements (seizures)
- Headache
- Loss of consciousness
- Chest pain or abdominal pain
- Swelling of the body
- Pain while urinating (Share.upmc.com 2021)

If fever is not treated quickly, it means that we are ignoring the symptoms of the underlying causes and giving time for the infection to flare up or worsen.

Treatment

Initial measures to lower temperature can be tried without medication. These simple steps are:

- Remove heavy clothing, open windows, and keep the environment cool
- Drink plenty of fluids
- Obtain adequate rest

However, never use cold baths to bring the fever down. Also, do not cover the patient up with multiple blankets for the chills and shivers. Instead, try to keep the environment cool.

Give acetaminophen or paracetamol to lower the temperature. These are over-the-counter drugs and are easily available in pharmacy stores without a prescription. Follow the instructions on the medication to take as directed.

Some other natural ways to bring down a fever are:

- Repeated use of a sponge or towel that has been soaked in clean, cold water and partially squeezed to clean the forehead, chest, groin, and wrist (this helps bring down the body temperature)
- Lukewarm bath or shower (this restores energy)
- Drinking lemon water or increasing vitamin C (this improves immunity)

- Eating nutritious soups or broths that contain herbs such as basil and spices like pepper (this can improve blood flow, help the body cool by producing sweat, and provide the essential energy needed to fight the infection, thus bringing down the body's temperature) (M 2021)

Mouth Sore

Mouth sores occur due to ulcer formation. Mouth ulcers are painful sores in the mouth. Mouth ulcers are also known as canker sores or aphthous ulcers.

The causes of mouth ulcers are:

- Biting the inside of your mouth
- Braces or dentures that are too big or too small
- Brushing your teeth too hard
- Burns from hot drinks
- Hormonal changes
- Low vitamin B12 levels
- Low iron levels
- Genes
- Smoking
- Stress
- Weakened immune system (the body's defense to fight against germs)

Mouth ulcers are:

- Wound/s with a red border and white center
- Occuring inside the mouth on the lips, cheeks, tongue, and roof of the mouth
- Painful and burn whether eating or not (Tagazier 2021)

Treatment

Most mouth sores usually resolve on their own, even without treatment. It may take up to 7 to 10 days. If you want to treat them during that time, you can take:

- Pain medications in the form of oral supplements, as directed, or use mouth rinse (3 times per day for 7 days).
- Vitamin or iron supplements if you have decreased levels of these.
- Anti-microbial mouthwash or gel (3 times a day for 7-10 days).

Other home remedies that can be tried to heal mouth sores are:

- Rinsing the mouth with clean, lukewarm saltwater.
- Using ice cubes to numb the sores. Keep small chips of ice on the sore and let these dissolve slowly.
- Drinking cold fluids instead of hot food or fluids.
- Avoiding spicy food.
- Brushing your teeth gently. Use a soft-bristle toothbrush.

- Applying a thin paste of baking soda and water to the blister (MedicineNet 2022).

These home remedies can be used 3-4 times per day until the mouth sores heal.

Sprains

A ligament is a short, tough band of tissue that connects two bones and holds a joint together. If a wrench (pull) or twist of the ligament occurs, it is known as a sprain. The sprain causes pain and swelling in the joint and limits movement. Sprains commonly occur in the wrist and ankle.

If you sustain an injury resulting in a sprain, make sure to give adequate rest to the injured part so that you do not injure it further. There are no emergency "fixes" available. However, the management can significantly decrease the amount of healing time. Other measures that can be adopted in treating sprains are to:

- Apply a cold compression using ice cubes wrapped up in a cloth. The cold compression should be applied the first 2 days of injury, for 15-20 minutes, every 2-3 hours.
- Elevate the injured part for 2-3 hours/day until the swelling decreases, if it can be done safely and you do not suspect a fracture. If a fracture is suspected, you can go for a splint or sling (see pgs. 106, 111 for details).

- Use painkillers for pain relief as needed or 3 times/day (Brouhard 2022).

R.I.C.E.—you can refer to pg. 103 for details.

Dizziness

Dizziness is the condition in which you feel you are about to faint. The word "vertigo" is often used interchangeably with dizziness. However, vertigo refers to the condition where you feel that your entire surroundings are spinning (Medlineplus. gov n.d.).

Severe dizziness and vertigo can be handicapping to the individual as the person cannot work or perform his or her daily activities. There may be accompanying nausea and vomiting during the episodes.

Most cases of dizziness resolve on their own. However, you can give anti-vertigo medicines and anti-vomiting medicines (if there is accompanying nausea and vomiting). Anti-vertigo medicines are not available as over-the-counter drugs but anti-vomiting medicines are. In case of excessive vomiting, make sure to maintain adequate hydration by giving fluids.

There is also a go-to exercise you can try to remedy vertigo called the Epley maneuver. The patient sits in an upright manner with legs stretched out and a pillow (or something to lay their head on) behind them on a flat surface such as the floor.

Then the patient turns their head 45 degrees to the right and quickly lies down so their head is now on the pillow. This position is held, which includes the head being at a 45-degree angle, for 30 seconds. Then the patient slowly turns their head 90 degrees to the left without lifting their neck. After that, the patient turns the rest of their body to the left so that they are completely turned on their left side. Finally, the patient slowly returns to their original position, sitting up and looking forward. You can assist your patient by guiding their head accordingly depending on the step. This exercise can be repeated three times in a row. Note: your patient may feel dizzy the whole time (Watson 2019).

Epley Maneuver

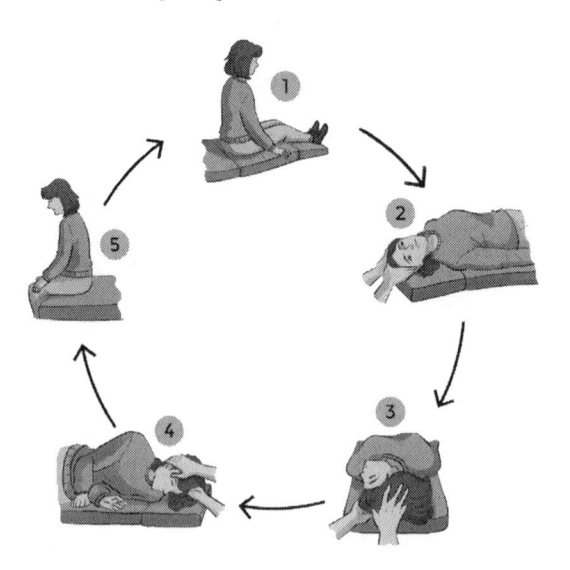

Takeaway

In this chapter, we have discussed the scenarios that are encountered frequently but which can be managed on an individual basis in most patients without requiring professional medical attention. Only some patients report severe symptoms and require hospital care. Also, we can manage the scenarios without having a lot of materials or resources around us. Thus, learning how to diagnose these conditions and their treatment can be beneficial on a day-to-day basis.

7
HANDLING SEVERE HEALTH EMERGENCIES

Throughout this book, we have been talking about unexpected emergencies. We have discussed how to approach a patient and perform certain life-saving maneuvers. In this chapter, we will take an individualized approach to certain specific scenarios. These are high-risk situations in which your prompt action can prove to be life-saving to an individual.

Let us now deal with these scenarios one by one.

Shock

Shock is the decrease of oxygen supply to the brain as a result of poor circulation. This decreased circulation in turn results in sub-optimal functioning of vital organs in the body (like the brain, kidney, heart, etc.). The causative factors can be divided into the following categories:

- Due to heart problems (cardiogenic shock): As in heart attack or cardiac arrest (sudden stopping of the beating of the heart)
- Due to very low volume of blood (hypovolemic shock): Excessive bleeding, dehydration due to severe diarrhea, and vomiting
- Due to infections (septic shock): Severe bacterial and viral infections
- Due to severe allergic reactions (anaphylactic shock): Insect bites, stings
- Due to damage of the nervous system (neurogenic shock): Damage to the brain or spinal cord (Medlineplus.gov n.d.)

Handling Severe Health Emergencies

Shock can proceed through various stages before death occurs.

First is the compensated phase. Whenever any imbalance occurs in our body, the initial response is to try to compensate for the deficit. Thus, when there is decreased circulation, the body tries to compensate for it by increasing the heart rate and thus maintaining blood pressure. If the factor causing shock is treated at this time, the outcome is good.

However, if the causative factor is still not corrected, the second stage ensues. This stage is also known as the progressive or decompensated phase. In this stage, the blood pressure drops suddenly, resulting in decreased oxygen supply to the organs. If this occurs for a long period, within minutes, irreversible damage will have occurred and the patient may not improve, even with aggressive treatment (Forgey 2020).

In any severe illnesses or accidents, consider the possibility of shock. Look for signs of adequate circulation by feeling the wrist or the carotid (neck) pulse (see pg. 52). If you feel a weak and rapid pulse (>140/min in adults and >180/min in children), deduce that the body is trying to compensate, meaning you should actively intervene to prevent it from progressing to the second stage. You can also listen to the bare chest. If you do not find a pulse or hear the heartbeat, start CPR (see pg. 116) immediately. After the patient stabilizes, proceed to treat the underlying cause.

If the shock has occurred due to severe allergic reactions, administer epinephrine immediately in his or her thigh. If a severe infection is a cause, try to determine the source of infection, if possible, then start antibiotics accordingly. If excessive blood loss has occurred, if available, administer fluids (saline—bottled salt water) either via mouth or veins, assuming blood transfusion is unavailable, and try to control the bleeding (see pg. 169).

Poisoning

Poisoning is injury or death caused by exposure to harmful substances, either by ingestion (eating or drinking), breathing in, touching, or injection into the blood. Various medicines, if taken above the required doses, also act as poisonous substances. Other examples are detergents and carbon monoxide, which is released when coal burns. Poisoning can be either accidental or intentional.

We can suspect poisoning in a person if any of the following symptoms occur:

- Burning or redness around the mouth area
- Vomiting
- Difficulty breathing
- Tiredness
- Confusion
- Breath that smells like gasoline or paint thinners

Immediately call for help, if possible, if the above signs are seen in a patient. If help is not available, begin first aid immediately.

Remove the remaining poisonous substance from the mouth, if visible. If the person has breathed in the poisonous substance, shift the person to fresh air immediately. If the substance has come in contact with the skin, remove the clothes and wear gloves or some other protective gear on your hands to avoid exposure. Then, rinse the skin with water for 15 to 20 minutes. If eyes are in contact with the substance, rinse them with cool or lukewarm water for 15 to 20 minutes.

If the person vomits, turn the person to his or her side so that the particles do not enter the airway.

If the person does not cough, vomit, or respond, i.e., he or she is unconscious and you suspect that he or she is not breathing and has no pulse, start CPR immediately (see pg. 116).

If you find the container of the poison with warning labels and a guide on what to do in case of poisoning, act accordingly (Mayo Clinic Staff 2022b).

Bleeding

When someone sustains an injury due to cuts, bleeding almost always occurs from the wound. It is essential to know the nature of the bleed for prompt management. Also, blood acts as a route for transmission of infection. Hence, first-aid is necessary for bleeding injuries.

The nature of the bleeding can be determined by the amount and color of the blood. If there is a trickling of blood that subsides on its own after some time, it is most probably a capillary bleed. Capillaries are the smallest vessels through which the blood flows. Venous bleeds are darker in color and ooze out of the wound. Arteries carry bright red blood. If an injury to the artery occurs, the blood spurts out from the wound.

Bleeding, if it occurs for a prolonged period in a significant amount, can result in hypovolemic shock (shock due to decreased amount of blood) (see pg. 166) and death. Hence, bleeding needs to be managed promptly. If a person is bleeding profusely but has a pulse, try all measures to decrease bleeding. If the person is bleeding and does not have a pulse, start CPR immediately.

The following measures can be taken to decrease bleeding from the wound:

- Put on gloves before you handle a patient who is bleeding. It will help prevent diseases that transmit through blood like HIV and Hepatitis B.
- Apply direct pressure on the wound with gauze pieces. If the previous ones are soaked, add more layers rather than replacing them. Removing them will interfere with the blood clotting process.
- Also, note that spraying Afrin or Neo-Synephrine nasal spray onto the gauze that is then placed on the skin has been found to help stop bleeding. It is believed that Afrin and sprays like it contain blood vessel constrictors (Heimark 2010).

- After the bleeding stops, cover the wound with a clean bandage (Brouhard 2022a).

Chemical Wounds

Chemical wounds are caused by strong substances (acids, cleaners, gasoline) which result in corrosion of the skin or the inner lining of our organs.

There are 3 modes by which the chemicals can enter our body:

- Contact with skin
- Ingestion by mouth
- Inhalation

Symptoms differ according to the organs which have been exposed.

If a chemical burn has occurred to the skin, the skin may appear red, swollen, and painful. Blisters (painful swelling on the surface of the skin containing clear fluid) may develop. In severe cases, the skin appears black and charred. If the burn is severe enough to damage vessels, the area may be numb.

If the chemical gets into the eye, the patient will have pain and redness in the eye, with decreased vision.

If the patient has breathed in the chemical, he or she will have a cough with or without difficulty breathing.

Family Survival Medicine Handbook

If the patient has eaten or drunk the substance, he or she will have pain and a burning sensation throughout the gut. He or she may experience nausea and vomiting multiple times (Intermountain Healthcare n.d.).

Prevention is of utmost importance in this condition. Keep chemicals out of reach of children. Do not keep chemicals in water bottles, unless clearly labeled. Do not mix chemicals, as the fumes released can be poisonous. When using chemicals, always use gloves and also make sure you protect your eyes. Do not let the chemicals come in contact with your clothing.

For minor burns, basic first aid is enough most of the time.

If you encounter a case of a chemical burn:

- If the chemical is a dry powder and any remains on the victim, after putting on gloves, brush off the remaining substance.
- Remove any garments or jewelry that touched the chemical.
- Rinse off the affected area for at least 20 minutes under running water.
- Painkillers (like ibuprofen and paracetamol) can be given.
- Do not try to remove skin or blisters.
- If the poison is inhaled, shift the patient to fresh air and administer oxygen if available.

Bandage the affected area and clean the wound once or twice

a day with cool water. The bandage should be put on loosely. Don't put pressure on the burnt skin (Mayo Clinic Staff 2022a).

Always remember to call for medical help before you start administering first aid in moderate to severe cases and in people who have inhaled or ingested the chemicals.

Concussions

When a trauma (injury) occurs to the head, it can result in the loss of mental function for a short period of time. This is known as a concussion. The patient may have delayed or absent response due to temporary memory loss after the hit. They may also give an account of blurry or starry vision (NHS 2021).

However, it is extremely important to differentiate concussion injuries from other moderate or severe head injuries. If not, the consequences can be devastating.

Seek medical help immediately if the following signs are seen:

- Loss of consciousness
- Difficulty walking/loss of balance
- Slurred speech (incomprehensible speech)
- Persistent (continuous) vomiting
- Sudden deafness
- Clear fluid discharge from the nose and ears

Concussions can be addressed in the household by providing immediate first aid. You can:

- Apply a cold compression to the injured area. Wrap ice cubes in a clean towel. Leave the cold compress in place for 20-30 minutes every 2 to 4 hours.
- Give paracetamol (Tylenol) or acetaminophen for pain control. Do not give aspirin or NSAIDs (non-steroidal anti-inflammatory drugs) like ibuprofen as they can cause bleeding.
- Avoid alcohol or sleeping tablets (NHS Inform 2021).

Patients who have sustained a concussion should be observed for at least 48 hours after the injury to see if they develop danger signs (see pg. 173 for details). If so, surgical intervention may be required.

When Should You Return to Normal Activities?

The patient should return to work only when he or she feels completely recovered.

Wait for at least 3 weeks to return to contact activities like sports. However, this guidance differs and some organizations recommend a step-by-step approach which states that when the patient feels completely well, start with light exercises such as walking. Then, move on to more moderate and intense activities (NHS Inform 2021).

Burns

Burns constitute a medical emergency if a large area and depth have been involved. If a minor burn occurs, it can be managed at home. The first step is to remove the cause of the burn.

There are different types of burns, which include:

- Thermal burns (heat)
- Chemical burns
- Radiation burns (radioactive material exposure)
- Electrical burns (Webmd.com 2021)

The deeper the skin involved in the burn, the more severe the injury is considered. The severity of the burn can be classified as:

- First-degree burns: These only affect the skin's surface layer. It becomes red and swollen.
- Second-degree burns: These generate blisters and two layers of skin redness. If the burn is more than three inches wide, or if it is on the face, hands, genitalia, buttocks, feet, or covers a major joint (such as the shoulder or hip), it is a major burn (Brouhard 2022b).
- Third-degree burns: These are considered to be major burns and affect all layers of skin. The skin becomes blackened or whitened. If nerves are involved, the skin becomes numb (Brouhard 2022a).

If a person sustains a minor burn injury, do the following:

- Rinse the affected area with cool water. In case of chemical burns, remove any contaminated clothing after putting on your gloves.
- Apply a clean bandage. Do not put pressure on the wound.
- Give painkillers for pain relief.
- Do not burst any blisters that may have formed.

If a major burn is sustained, call for emergency assistance and treat as follows:

- Assess for ABCD if the patient starts to show signs of shock or abnormal breathing.
- Remove any restrictive clothing or accessories near or on the burned area.
- Lift or elevate the burned area above the patient's heart if possible.

Apply a light gauze or clean cloth loosely (Mayoclinic.org 2022).

Electrocution

Whenever a person comes into contact with a naked electrical live wire, the current passes through his or her body. This is known as electrocution. Before beginning with the management of electrocution, let's go through the basics of electric current.

There are microscopic particles in a substance that carry negative charges. These particles are known as electrons. Since an electron carries a negative charge, it is always attracted to positive charges and tends to flow towards them. This presence and flow of charges are known as electric currents. When a path is provided for the current to flow—for example, through wires and batteries—the path is known as an electric circuit (Electronics-notes.com n.d.).

Electrocution can be either mild or severe and can also lead to death, depending upon the intensity of the electricity and the period of exposure. The injuries can include burns on the skin which are visible externally, or there can even be damage to the internal organs, diagnosed only by proper medical examination.

In all cases of electrocution, the first step is to cut off the power source, if possible. If not, then remove contact between the patient and the source. Use insulating materials like rubber gloves, folded dry garments, or rolled-up newspapers for this purpose. You can also use a dry stick to cut off the patient from the source. NEVER use your bare hands or try cutting a naked wire with blades or scissors.

After the person has been detached from the source, call for medical help, if possible. Then, look for the signs of breathing and circulation. If the signs are absent, start CPR immediately (see pg. 116). Do not move a patient unless there is immediate danger, as the electric current may have caused damage to the spinal cord (Thebetterindia.com 2018).

If there is a visible burn on the skin, rinse the area with water and apply a loose bandage to the area.

Lightning

Lightning Safety Tips

The safest place to be when a lightning storm strikes is inside and away from:

- Water (i.e., showers, baths, and washing dishes), since lightning can pass through the plumbing system.
- Electrical equipment (i.e., do not touch computers, laptops, outlets, washers, dryers, and everything else that is connected to electricity), since lightning can pass through electrical systems.
- Windows, porches, doors, and anything made from concrete, since lightning can go through metal wires or bars in concrete walls or floors.
- Corded phones (cordless or cell phones are safe).

While it is safer to be inside during lightning strikes, if you happen to be caught outside, here are some tips:

- Move quickly away from higher ground (i.e., hills, peaks, and mountain ridges).
- Do not lie flat like a pancake on the ground! Instead, crouch down like you are squatting with your head tucked and hands over your ears. This creates as little contact with the ground as possible.

- Do not get under one random tree out in the open. If multiple trees are near, then get under the lower ones.
- Do not seek shelter on a cliff or overhang.
- Get as far away from all bodies of water (e.g., lakes and ponds) as possible.
- Move away from things that allow electricity to transfer, such as barbed wire, fences, power lines, and windmills.
- If swimming out in open water, immediately get to shore as quickly as possible.
- If you are out in open water in a boat, get into the cabin. If one is not available, drop the anchor as low as possible.
- Do not stay in open moving objects such as convertibles, motorcycles, or golf carts.
- Stay away from open spaces like beaches, golf courses, and parks.
- Steer clear of tall objects like trees and telephone poles because lightning tends to strike the tallest object nearby.
- If you are in a group, split apart from one another so that if lightning strikes the ground, fewer people will be injured.

Lightning Medical Treatment

If someone is struck by lightning, if possible, first call 911. Then assess the situation. You want to make sure you and your patient are less likely to get struck (i.e., if in an open field, move inside or wait for the storm to pass). A person hit by lightning is safe to touch because they cannot pass on the electrical current. Also, unless they were thrown a long way or suffered a bad fall, it is very unlikely that they suffered a broken bone. If they are bleeding or look as though they broke a bone, do not move them and continue with consent, ABCDE, etc. as previously discussed (see pg. 40). If you do move them, after doing so, check their pulse and breathing since more times than not, lightning causes those it strikes to have heart attacks. If this is the case, begin CPR (see pg. 116) immediately. If the person is breathing normally, check for other injuries such as burns (see pg. 175) or shock (see pg. 166), and treat each injury accordingly. If the area is wet and cool, cover the ground your patient is on with a sheet, jacket, or blanket to keep your patient's body temperature from dropping too low (hypothermia; see pg. 214) (Centers for Disease Control and Prevention 2022).

Caring for Someone in Quarantine

This is the era of COVID-19. With newer mutations every few months, a lot of people are getting infected. When infected people are kept separate to prevent infection from transmitting to others, it is known as isolation. When people have come into contact with someone who may have COVID-19, they are kept separate until test results are known. This is called quarantine.

This period can be both physically and mentally challenging to the patient. And as much as you love your friends and family members who are infected and want to care for them a lot, there is also a great risk of getting infected yourself. Thus, for anyone who has been infected and is in isolation or quarantine, you need to care for them by following a few key measures:

- As much as possible, ensure that the caregiver is fully vaccinated and not suffering from long-term illnesses themselves.
- Always make sure to wash your hands with soap and water for 40 seconds and disinfect your hands with sanitizer.
- Then, put on a gown, gloves, and mask so that you do not spread the infection.
- Make sure the room of the patient is well ventilated.
- If possible, avoid a situation in which other members of the household use the same bathroom as the patient.
- Ask the patient to put on their mask when going to the bathroom and disinfect the toilet seat and door handle after use.
- Always keep a thermometer, saturation probe (an instrument that measures the blood oxygen levels), and necessary medications like paracetamol (Tylenol) or acetaminophen at the bedside of the patient (Unicef. org 2021).

However, these guidelines are constantly changing as new facts about COVID-19 are being discovered. So, please make sure to check the authentic sources from CDC (Centers for Disease Control) or WHO (World Health Organization) for updated quarantine and isolation protocol. The guidelines have even changed while this book was being written.

Takeaway:

In severe health emergencies, quick, proactive intervention is key to rescuing victims. However, calmness is also very much required in these scenarios. Hence, avoid making rash decisions or doing things that could make the situation worse. As previously stated, this is a lot of important information to be familiar with so we highly recommend taking courses in addition to reading this book so that if/when you are presented with this situation, you are calm and remember what to do to the point that, if needed, you can even teach someone else. Some would say that is the definition of a true expert.

8
HANDLING HEALTH-RELATED EMERGENCIES

Once, there was a family in one of our neighborhoods and some of us decided to go camping with them for 5 days. The dad of the family was an obese guy who was taking medicines for diabetes (increase in blood sugar levels) and high levels of cholesterol (fat that is harmful to the body). It was the third day when we were walking that he complained of sudden pain in the middle of the chest. He started sweating, turned pale, and in minutes fell to the ground and was unconscious. His gestures made us suspect that he had a heart attack. Thanks to the seminar on the heart attack that we had taken only a few weeks before, we gave him aspirin and he recovered and became conscious. We then took him to a nearby health facility.

The man had diabetes as well as high blood cholesterol, which put him at higher risk of having a heart attack. However, not everyone who has a heart attack is as lucky as him. The blockage of the vessels of the heart must have only been temporary, being that a single dose of aspirin improved his condition. Others who have a complete blockage of a major vessel tend to succumb to the condition.

This story must have already given you an insight into what we are going to discuss in this chapter. Certain emergencies that occur are related to the patient's history and previous health conditions. In some cases, these emergencies may occur for the first time, mostly in individuals who do not bother getting regular health checkups.

Now, let us discuss some of these health-related emergencies. We will start by discussing heart attacks in detail.

Heart Attack

We all know that the heart pumps blood throughout our bodies. But for this action, even the heart needs energy, which comes from the blood itself. The vessels through which oxygenated blood is delivered to the heart are named the coronary arteries. When a coronary artery gets blocked, the blood supply to the heart suddenly stops such that the heart is no longer able to pump blood throughout the body. Thus, the patient faints. This phenomenon takes place within a few seconds and the patient may lose his or her life. Hence, immediate first-aid action is required for this process to reverse.

The patient may complain of sudden pain or tightness in the center of the chest. The pain may also be felt to pierce up to the back. He or she may also experience sweating, difficulty breathing, and fainting.

The first-aid measures to follow in a patient whom you suspect to have suffered a heart attack are as follows:

- Call for emergency medical help. A heart attack is an extremely risky medical condition. So, make sure you leave no stones unturned. It is best for the patient to have professional help.
- If help is not available and you do not have alternative ways to take the patient to a hospital, give aspirin for the patient to chew and swallow. Aspirin prevents blood clots.
- If nitroglycerin (a medicine used to treat chest pain) is

available, either to you or the patient, ask the patient to keep it under their tongue or follow any other methods that the doctor may have prescribed. Nitroglycerin may be available if the patient has suffered a heart attack previously.
- If the person becomes unconscious, start CPR as described on pg. 116.
- If an automated external defibrillator (AED) is available and you are trained in using it, do make use of it as well (Mayo Clinic Staff 2022b).

Stroke

If a person suddenly cannot move their hand and foot on one side of their body, their speech becomes incomprehensible, and their mouth becomes deviated, all these signs may indicate a stroke.

The brain needs oxygen to function. If the blood vessels which carry blood to the brain clog or rupture, part of the brain becomes damaged. Hence, the above-mentioned signs and symptoms are seen in the patient. Sometimes, the patient may also have nausea and vomiting, dizziness, and loss of consciousness (Centers for Disease Control and Prevention 2022b).

If you suspect that you are having a stroke or if you are caring for others who have just suffered a stroke, keep the following points in mind. Time is of great essence in managing a case of stroke. In cases where the artery is blocked, it is possible to open the area within 4.5 hours. Hence, call 911 and immediately get the patient to a nearby facility. If you can achieve this, the outcome can dramatically improve and the patient will be thankful to you for their entire life.

However, if accessing a medical facility is not possible at all, keep the patient in a comfortable position, preferably on his or her side, so that if vomiting occurs, he or she does not choke. Do not allow the patient to eat or drink anything.

If they are not breathing (look for signs as discussed on pg. 40), start CPR immediately and continue until the patient resumes breathing (see pg. 116 for details).

Do not try to move the hands or feet of the patient if they are showing signs of weakness (McDermott 2019). It won't help!

Allergic Reactions

Allergies occur when a person develops an unusual response to any substance or germs present in the food or environment. There is always a system activated in the body that identifies which foreign substances are harmful. The foreign substance is termed an "antigen" and the defense produced by our body is known as an "antibody" (The Editors of Encyclopaedia Britannica n.d.-a; The Editors of Encyclopaedia Britannica n.d.-b). These antibodies are proteins. Proteins are large, complex molecules that are essential for the structure and function of the body. Protein is made up of several atoms (basic building blocks of all matter) that combine to form a molecule.

Body's Response to Germs: Antibodies

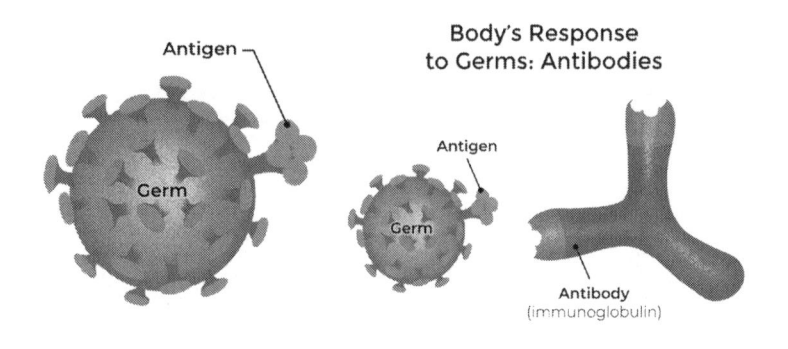

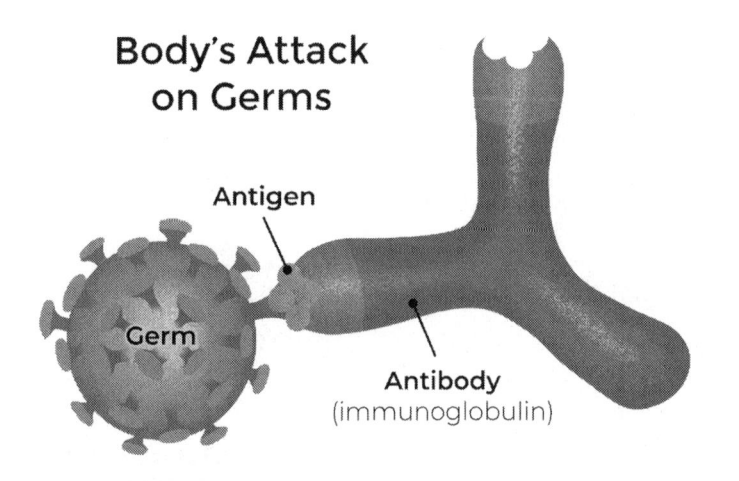

Antigen
Germ
Antigen
Germ
Antibody (immunoglobulin)

Body's Attack on Germs

Antigen
Germ
Antibody (immunoglobulin)

Now, it is possible for the body to identify foreign substances as harmful when in fact they are not, which causes several symptoms. In some, the symptoms may be mild, like coughing, sneezing, and watery eyes, whereas in others, the symptoms may be severe, like swelling of the face and the entire body and difficulty breathing. If such severe symptoms occur, there can also be signs of shock, like an increase in heart rate and decrease in blood pressure. This condition is known as anaphylaxis and needs to be managed immediately, as it can be life threatening.

Let us have a detailed overview of the symptoms that occur in different types of allergies (Pietrangelo 2020):

Symptoms	Environmental allergy	Food allergy	Insect sting allergy	Drug allergy
Sneezing	X	X		
Runny or stuffy nose	X			
Skin irritation (red, peeling, itchy)	X	X	X	X
Hives (itchy, raised patches of skin)		X	X	X
Rash		X	X	X
Trouble breathing			X	
Nausea or vomiting		X		
Diarrhea		X		
Shortness of breath or wheezing	X	X	X	X
Watery or bloodshot eyes	X			
Swelling around the face or area of contact			X	X
Rapid pulse			X	X
Dizziness			X	

The mainstay of treatment of any allergic reaction is to avoid the agent causing the reaction.

In environmental allergies, the most common causes are the pollen grains of plants, dust, and smoke. Since environmental allergies involve a response to everyday surroundings, unless the cause is known, it is recommended to wear a mask as much as possible to prevent these allergens from entering the nose or mouth. In case of the sudden onset of symptoms, have the patient take allergy medications such as Advil, available as an over-the-counter drug. Most anti-allergic medications cause drowsiness and should be taken before sleep. Decongestants are also recommended. They remove nose obstructions, making it easier to breathe. These decongestants are available in either oral or drop formulations. Take them as directed on the packaging. For natural decongestant remedies, refer to pages .

With food allergies, have the patient avoid eating food that triggers their allergic reaction. If you accidentally consumed something, allergy medications can help. These can be taken immediately as directed on the packaging.

Treatment of insect stings is discussed in detail on pg. 150.

If allergic symptoms occur immediately after taking a particular medication, stop taking it and, when available, ask the doctor to change it.

However, in case of a severe allergic reaction (anaphylaxis), the treatment approach differs. Contact 911 if possible, but whether help is on the way or not, look for an EpiPen. The person may have this if he or she has already suffered similar episodes in the past. Inject the pen on his or her thigh.

Make sure the individual is lying comfortably on their back under a blanket and loosen any restrictive garments. Turn the patient on their side if they are vomiting or bleeding from their mouth to prevent choking. If there are no signs of breathing or circulation (see pg. 40 for more details), start CPR (see pg. 116) immediately (Mayo Clinic Staff 2022a).

Seizures

Seizures are uncontrollable body movements occurring due to abnormal electrical activity in the brain. Seizures can be of different types. In some, abnormal body movements may only occur in a specific part of the body, whereas in others, they may occur across the entire body. In some, there may only be starting episodes (MedlinePlus n.d.).

Seizures are also called "convulsions." The cause of seizures may be unknown or there may be a mass or infection in the brain. Convulsions can also be triggered by flashes of light and stress.

The first-aid measures which we can administer to a patient with seizures are:

- Wait for the episode to pass. Call 911 if possible if a single episode lasts for more than 5 minutes.

- If the person has anything in their mouth, turn the person to the side to prevent choking.
- Roll the person to their side, because they may bite their tongue and cause bleeding from the mouth.
- Look if they are wearing any sort of medical bracelet.
- Clear the surrounding area of any sharp objects with which the person could hurt themselves.
- With the person lying down, put something soft like a jacket under their head so that they don't hurt themselves.
- Loosen the clothes around the neck and make them comfortable for the patient.
- But most of all, do not panic, and talk to the patient calmly until they regain consciousness (Centers for Disease Control and Prevention 2022a).

That said, during seizures, we may not be able to provide definite care for the patient but only try to help the patient until the seizure episode stops. Following this, the patient will require proper assessment, diagnosis, and treatment in a medical facility, when available.

Pregnancy: Helping Someone Give Birth

Delivery of a baby ideally requires a medical facility with medical equipment available. If possible, call for professional help or drive the woman to the nearest hospital.

The following section was written by a member of our team who successfully delivered her first child naturally, without medication and without complications, by utilizing hypnobirthing. As she describes it, the midwives and nurses were bewildered at how calm both she and her husband were. Disclaimer: she walked at least two hours, three to five times a week, from the start of her pregnancy, and that is also believed to have been very beneficial. That said, the following techniques and practices capture the key birthing practices she learned. We highly recommend that families who are going through the birthing process take a hypnobirthing class and study the book. If disaster strikes and hospital birth is no longer an option, you may be amazed at how beautiful your birthing experience can still be.

If time permits, go over these principles before a woman goes into labor and repeat them as much as possible. If there is not enough time, try your best to teach these laws while she is going through labor and delivery.

Rule No. 1: Help the mom relax

It is a blessing that women can give birth. It is also extremely important that women understand their worth. They were "designed" in such a way that they can give birth. The miracle and blessing of motherhood should never be overcome with false narratives that cause women to fear this experience. We can boost a pregnant woman's morale by understanding some laws (Mongan 2018):

- The Law of Psycho-Physical Response
- The Law of Repetition
- The Law of Harmonious Attraction

- The Law of Motivation

Although these laws may seem complicated, what they are trying to say is that you have to think positively and your body will respond with the desired outcome. How you think affects how your body responds. Positive thoughts cause your body to respond by producing positive hormones, which give a feeling of well-being.

It is extremely important to use these laws in favor of the woman's body during the entire process of childbirth. Thinking positive thoughts and reinforcing them with constant repetition in mind and speech enhances the body's physical capabilities positively.

Breathing Techniques

Breathing techniques are very effective to ensure calmness of the mind. Three types of breathing techniques are mostly used:

Calm breathing: Our bodies are under stress most of the time. To relax our minds, we can perform calm breathing in day-to-day life. It ensures that oxygen is delivered to every organ of our body. It is this oxygen that is necessary for the proper and smooth functioning of our bodies. For this, have the patient relax comfortably in a chair. They should breathe in through their nose and mentally count up to 4, then exhale, again through their nose, to a count of 8. Tell them to breathe and release all tension, to let go! If a woman needs to breathe through her mouth so as not to feel like she is choking or gasping for air, that is fine. The main goal is relaxation.

Surge breathing: The uterus (womb) undergoes a wave of contractions and relaxation, i.e., gradually opening and closing to open up the cervix (the gateway to birth). Women have been known to also describe contractions as a time when their stomach went through cycles of tightening and then relaxing. Surge breathing corresponds to this rhythm and assists in the process of labor. For this, the woman should be in a partially sitting position with both hands cupped on the abdomen. Have her slowly breathe in and count to 20, pause, and breathe out as slowly as before. She should feel her abdomen completely expanding like a balloon. By using this breathing technique, her body also practices the rhythm of contraction of the uterus.

Birth breathing: The third type of breathing is birth breathing. As the name suggests, this technique should be adopted during the process of childbirth. After the cervix opens enough to allow the baby to pass, the woman feels the urge to breathe downwards. It has been described as the same sensation one feels when one is going to poop. In fact, that is a great time to practice this type of breathing. Use the same technique as calm breathing but this time, the energy should be directed towards the back of the throat and down the spine. The woman should relax her anus and vagina during the entire process and see how her body easily accepts this process (Mongan 2018).

To Push or Not to Push During Delivery

Pushing the baby with all of the mother's strength is not considered to be the best method for childbirth, although it is being practiced worldwide. Pushing hard deprives the womb and the baby of the needed oxygen, as the mother is holding her breath. As we all know, the body needs oxygen to have enough energy to function at its best, thus easing the delivery process. Also, pushing hard removes her concentration from the positive thoughts and breathing exercises. However, the mother may alternatively be allowed to take up different positions according to her comfort so that the natural "fetal-expelling reflex" occurs smoothly. Put simply, the body does not need help pushing out a child.

Some natural positions in which to labor and deliver are:

- Standing upright: The partner or object supports the mother and bears some of her weight. This is also known as the "slow dance position."

Handling Health-Related Emergencies

Standing Upright Examples

- Squatting: Opens up the bottom of the pelvis.

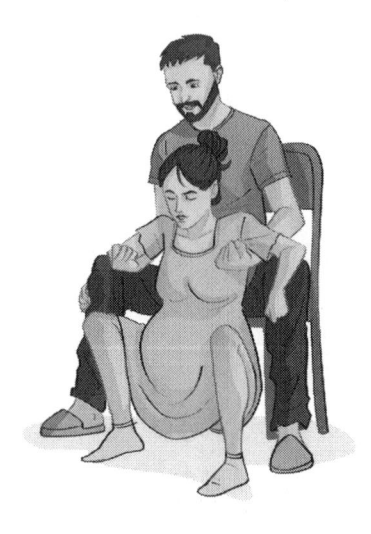

Squatting Position Example

Squatting Position Example

Handling Health-Related Emergencies

- Kneeling: Kneeling allows the mother to lean forward and tilt the hip bone (pelvis), allowing the baby to rotate.

Kneeling Position Example

- Raising one leg: Raising one leg on a chair stretches the pelvis on one side. This is best for babies in the wrong position.

Raise One Leg/Lunge Position Example

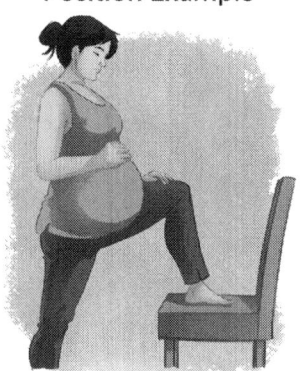

- Side-lying position: This helps correct the insertion of the head into the birth canal and reduces the likelihood of injury to the baby. The partner can support the leg and the back of the mother. Or a mother can support her own body while holding her bent knee, whichever is feasible and comfortable.

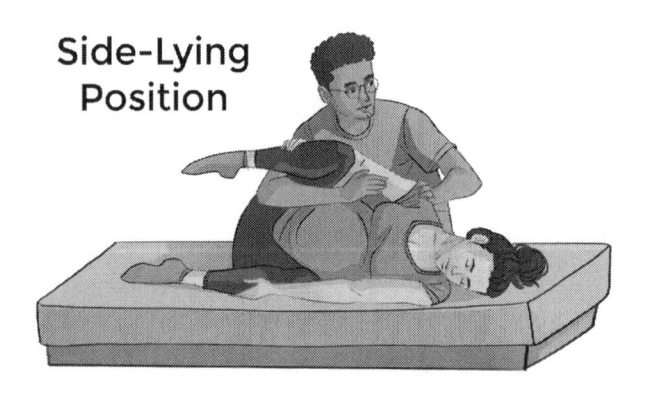

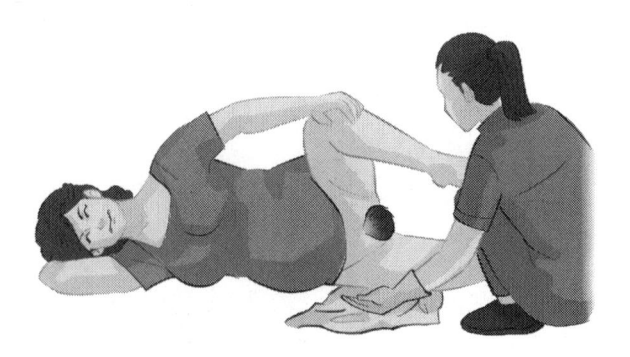

Side-Lying Position

- Semi-sitting: In this position, the partner supports you by sitting behind you and holding your body weight.

Handling Health-Related Emergencies

If your partner is unavailable, a bed or any other firm object can be used to support your back, which thus holds your body weight (Maximizedchiro.com n.d.).

Semi-Sitting Position Example

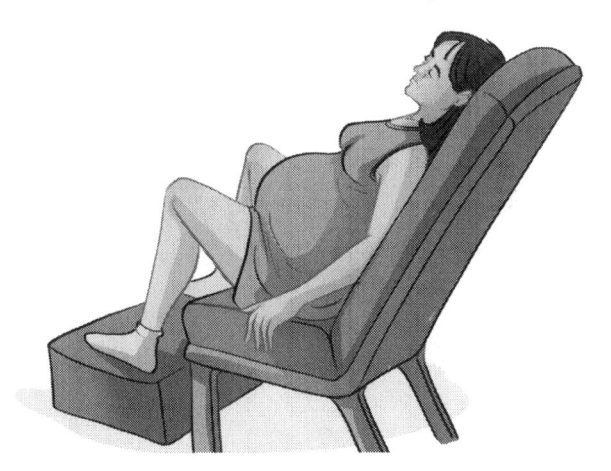

These positions are to be chosen according to the comfort of the mother and not for deliberate pushing out of the baby.

What Should I Do in a Case of an Unassisted Baby Delivery?

If you are expecting an unassisted baby delivery, gather the following items to ensure cleanliness and comfort to the mother.

- Towels, plastic sheets, and newspaper
- Soft blanket for the baby
- Gloves, if available

- Thick string, or sterile tape to tie off the umbilical cord
- Plastic bag for the placenta

Now that we know how to be mentally prepared during pregnancy and childbirth, the breathing exercises and positions that can help, and items needed to assist with delivery, let us learn the stages of childbirth, also known as labor.

There are 3 stages of labor

- The first stage is when the contractions start. The body, via contractions, signals that it is ready to start the delivery process. From the beginning of contractions to when the cervix (outlet of the womb) fully dilates (opens) is the first stage of labor. As you may have guessed, this is a good time for the lady to practice calm and surge breathing techniques.
- The second stage is the entire process of delivery of the baby. During this phase, it is good to practice birth breathing.
- The third stage is the stage after the delivery of the baby up to the delivery of the placenta, which is the organ that provides nourishment to the baby (National Women's Health n.d.). During this phase, it is good to still practice surge and calm breathing as the body is working to "birth" the placenta.

In the next illustration, the first four pictures represent the first stage of labor and the fifth picture represents the second stage of labor.

Phases of Delivery

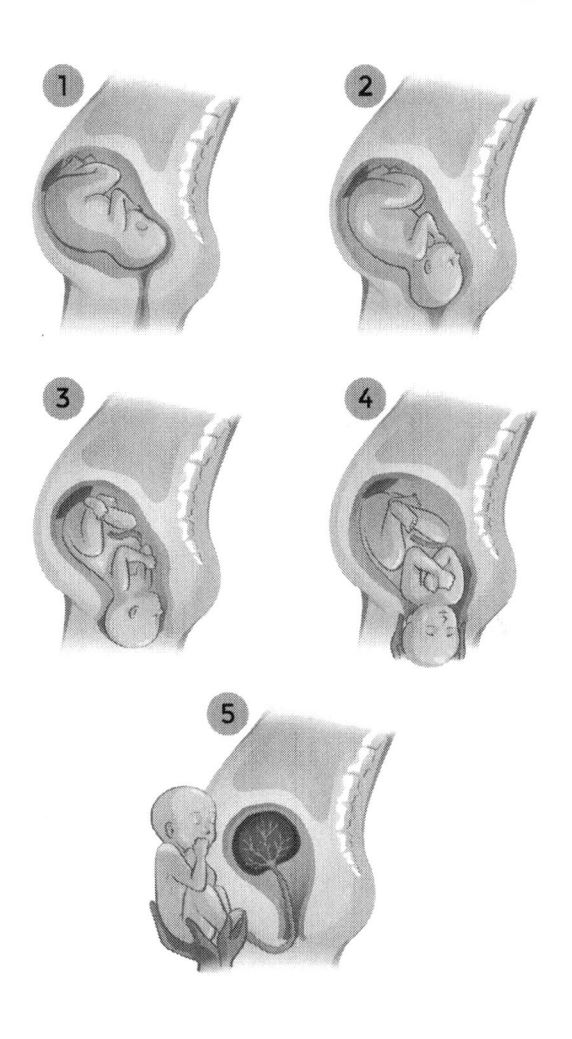

After childbirth, note the time of birth. Don't forget to congratulate the mother!

Then, hold the baby with the head positioned slightly lower than the body. This helps in draining the fluid from the nose and mouth.

Post Delivery

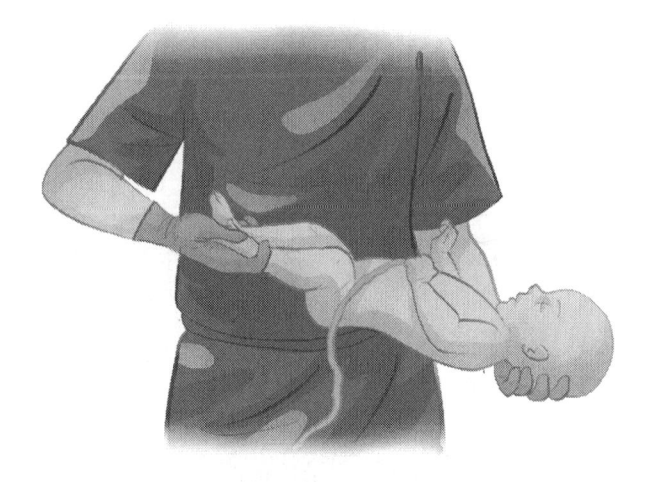

Gently dry off and wrap the baby with a towel. If the baby starts crying, it means that he or she is breathing. If not and the baby is blue, start CPR immediately (see pg. 116. for details).

Cutting the Cord

After the baby is born, wait until the cord stops pulsating. Then, tie off the cord with a thick string or sterile tape 2 to 3 cm from the newborn's abdominal wall. Tie a second knot 2 to 4 cm from the first knot. Then, cut between the two ties. Next, wrap the baby in a dry blanket and place the child on the mother's belly (Health.harvard.edu 2005; Clinicaltrials.gov 2021).

How to Cut the Umbilical Cord

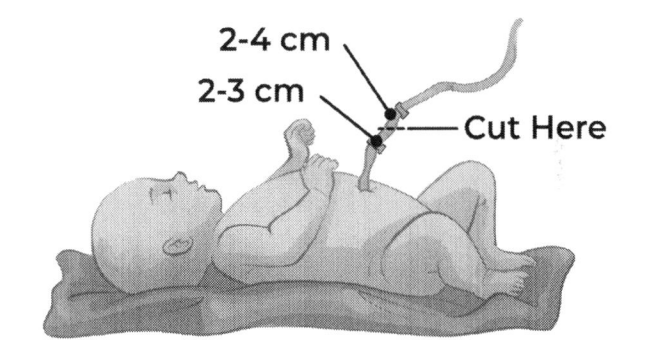

2-4 cm
2-3 cm
Cut Here

Takeaway

All of the scenarios discussed in this chapter are extremely critical. Time is of utmost importance in managing these conditions. Thus, in cases with previous health conditions and sudden deterioration, be sure to act fast so that you can do your best to save the life of the patient. Also, when knowingly entering a wilderness situation, if possible, everyone should know the others' current and past major medical conditions. This is a great way to specifically prepare and be less caught off guard in the event of an emergency. Knowing this information could also lead to the quick action needed to save a life or stop things quickly from getting worse.

9
ENVIRONMENTAL EMERGENCIES

The environment can surprise us at times. It may be too cold or it may be too hot. Sometimes, it can rain heavily, leading to flooding of the area. At other times, it may snow so much that the roads are blocked and daily life is hampered to a great extent. If you are outdoors and not prepared for what may come, your body may not adapt to the extreme weather. This leads to the development of environmental emergencies.

A similar occurrence once happened with one of our friends. They were trekking into a lake situated at a very high altitude. Though the weather broadcast did not mention any adverse conditions, it snowed heavily at that height. It wasn't possible to move forward or return. Thus, they were left fending for themselves for the entire two days. Naturally, their estimation of resources was out of proportion and the trip was not as smooth as they had dreamt of; however, they safely made it back, thanks to the fact that the heavy snowing only lasted for two days.

These sorts of environmental emergencies may happen to anyone. Before dwelling on the conditions that might occur, let us go into some of the fundamentals of how heat transfer occurs in our bodies.

Heat always flows from a higher temperature to a lower temperature. This stands true not only for persons or objects but also for the environment. You feel hot or cold when your body is not able to maintain a constant temperature. If your body temperature is higher than the surrounding temperature, heat from your body is transferred to the environment as per the law of thermodynamics (heat principle). Thus, you start feeling cold, as you are losing heat. A similar principle occurs when your body temperature is lower than the surrounding temperature, i.e., you gain heat from the environment and thus feel hot.

Now, let us start with discussing the scenarios by which the environment can affect our bodies, one by one.

Sunburn

Sunburn, also known as sun poisoning, occurs when excessively exposed to heat. It might occur if you stay out on the beach for too long, go fishing, or simply work out a lot in the yard. It mainly occurs in the summer season. Sunburn can manifest in mild or severe forms.

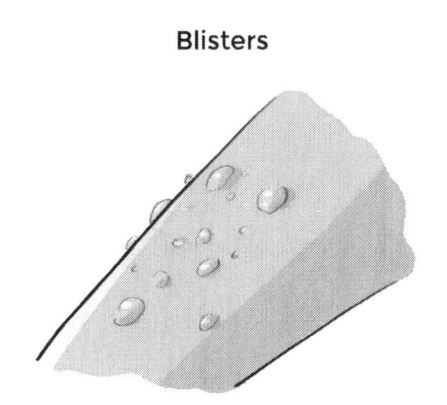

Blisters

Mild forms present as simple redness of the skin with blistering (raised area(s) of skin filled with clear fluid). The blisters may be small and noticed only when you try to peel off the skin with your hand.

There can also be flu-like symptoms like fever, nausea, and vomiting. However, these symptoms do not present immediately after sun exposure. The skin turns red after about 2 to 6 hours of exposure and peak effects are noted at 12 to 24 hours post exposure.

More severe forms present as large blisters which can be fluid-filled. There is a more severe form of redness and decreased water content of the body (dehydration), and infection may occur. Sometimes, it can also lead to shock with an increased heart rate and decreased blood pressure (see pg. 166 for details on shock) (Cunha 2021).

As sunburn occurs due to heat, if you or someone you are with experiences any of the above symptoms, apply the following actions:

- The first step is to cool off the skin. You can do this with damp clothes or by taking a cool bath.
- Avoid more sun exposure.
- Apply aloe vera gel, hydrocortisone cream, or calamine lotion to the affected area three times a day until the burning sensation decreases and the color of the skin returns to normal.
- Take painkillers like ibuprofen or paracetamol as directed on the bottle until the pain diminishes.
- Drink enough water to substitute for the sweat lost from your body.
- Don't break the blisters. However, if one ruptures on its own, apply antibiotic cream to prevent infection (Mayo Clinic Staff 2022b).

Heat Exhaustion

When someone is overexposed to extreme high temperatures, they get heat exhaustion. It is a mild kind of heat-related disease. The body sweats out its water and salt content when it is subjected to intense heat. It is potentially fatal when heat sickness develops into heat exhaustion and finally heat stroke. Thus, heat exhaustion should be treated from the very early stage when heat cramps occur.

The other symptoms with which a person can present are:

- Dizziness
- Nausea and vomiting
- Dark urine
- Muscle tightness or cramps
- Headache
- Pale skin
- Excessive sweating
- Rapid heart rate
- Confusion
- Dry mouth and lips (Ansorge 2020)

If you believe a person has suffered heat exhaustion, start measures to cool the person down and reverse the process. These can include:

- Asking the person to rest.
- Removing or loosening the person's clothes.
- Giving oral rehydration salt (ORS)—This is salt, sug-

ar, and water available in preparation with additional nutrients, but for emergency purposes, it can also be self-made. One sachet of ORS is dissolved in one liter of water and should be used within 24 hours. This will help to replenish the fluids and minerals (substances present in the food required for normal function like calcium and potassium) lost from the body. If not available, give plain water.
- Wrapping the person in wet clothes and fanning the patient. It will help them to cool down (Skinsight.com n.d.-b; Sja.org.uk 2021).

Take the person's temperature at the beginning before starting treatment. Then, start treatment and record their temperature every few minutes. You can stop using cooling methods once the temperature falls below 100 degrees Fahrenheit, but then for the next three to four hours, check the person's temperature every 30 minutes to make sure it stays low (Skinsight.com n.d.-b). For a description of how to measure the body temperature, see pg. 70.

Heatstroke

The most serious heat-related sickness is heatstroke. If it is not treated right away, it might be fatal. The core body temperature rises as a result of the body's inability to release extra heat. The interior body temperature is another way to refer to core body temperature. See pg. 70 for how to measure internal body temperature.

Environmental Emergencies

If heatstroke persists, the person often loses the desire to drink and the next stage is fainting and even death. Thus, immediate care for the person is vital.

The person may have the following symptoms:

- Fever of 104 degrees Fahrenheit (40 degrees Celsius) or higher
- Changes in mental state, such as agitation, disorientation, or slurred speech
- Hot, dry skin or excessive perspiration
- Nausea and vomiting
- Pale skin
- Rapid heartbeat
- Fast breathing
- Headache
- Fainting

As this is an emergency condition, if you suspect heatstroke, call for professional medical help immediately. Then, the following measures should be taken:

- Place the person in a cool tub of water or melting ice water.
- Sponge with cool water.
- Fan the person.
- Cover with cool clothes.
- If the person is conscious and oriented, give them fluids to drink: ORS or plain water (if ORS is not available). One sachet of ORS is dissolved in one liter of

water and should be used within 24 hours. Encourage the patient to drink frequently.

Follow these measures until the person shows signs of improvement or their temperature returns to normal.

However, if the person suddenly shows no signs of breathing or circulation (see pg. 40) or shows signs of difficulty breathing, start CPR immediately. See pg. 116 on how to perform CPR (Mayo Clinic Staff 2022a).

Hypothermia

When the body is exposed to excessive cold and cannot maintain its internal temperature, hypothermia ensues. The body responds to this by shivering and increasing the normal functions of the body to produce heat. The skin turns bluish. Also, blood vessels narrow to preserve heat. Depending upon the health condition of the patient, hypothermia may occur within minutes to hours of exposure to excessive cold. Patients with previous health conditions and poor nutritional status are more likely to easily suffer from hypothermia.

If the body temperature is less than 94 degrees Fahrenheit, it is classified as hypothermia. If the body temperature is less than 90 degrees Fahrenheit, it may disturb the rhythm of the heart and can also result in cardiac arrest (stopping the beating of the heart). If it is less than 86 degrees Fahrenheit, check for a pulse (see pg. 52). If absent, start CPR immediately (see pg. 116 for how to perform CPR) (Wildmedcenter.com 2016).

If you suspect a person is suffering from hypothermia, provide the following first-aid measures:

- If the person is wearing wet clothes, remove them and add an insulating layer like a reflective blanket or add clothing in layers.
- Seek shelter for the person and provide him or her with blankets.
- Administer heat packs to the person's armpits, sides of the neck, and groin (the part between the thighs and the abdomen). See the next illustration.
- Due to shivering, the person may have lost a significant number of calories. Give the person something to eat. Start with simple carbs like sugar or fruit juice and then you can switch to protein and fats like milk and eggs if they are not nauseous (feeling sick with an inclination to vomit) or vomiting.
- The cold may have induced frequent urination. Give the person fluids and salts to replace the loss.

Applying Heat Packs for Treatment of Hypothermia

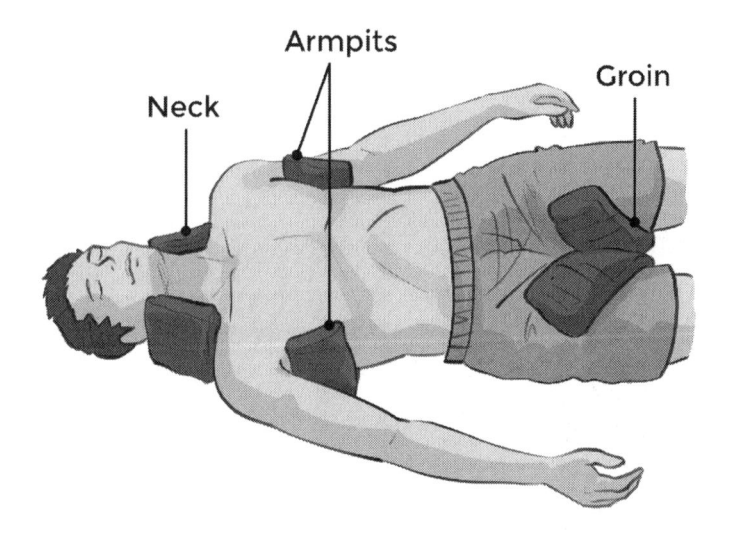

All these are measures for mild cases of hypothermia.

In moderate (below 90 degrees F) to severe cases (below 86 degrees F):

- Handle with extreme care as when the core temperature drops below 90 degrees Fahrenheit, the heart is likely to have an abnormal rhythm and can also go into cardiac arrest (no heartbeat).
- Food or fluids should not be taken by mouth.
- Rapid rewarming should not be undertaken, as this can also increase the chance of cardiac arrest.

- If the core temperature has fallen below 86 degrees F, begin CPR and rewarming until the core temperature reaches 86 degrees F (Wildmedcenter.com 2016).

Also monitor the body temperature every few minutes in order to treat accordingly. For details on how to measure body temperature, see pg. 70.

Frostbite

Frostbite occurs when the body or a body part is exposed to freezing temperatures. This is a serious medical emergency. The exposed skin first becomes red and painful, or white and numb. Then, blistering (swelling with clear fluid inside) occurs in the skin. The skin becomes hard due to the freezing of the blood vessels. When the body becomes exposed to extreme levels of cold, the blood vessels in the periphery (hands and feet) become so constricted (narrow) that the blood supply to these parts is cut off. This leads to the blackening and death of tissue which is known as gangrene (Marks 2021).

Frostbite is a rare phenomenon but can occur in people who go on outdoor activities, live or travel in cold climates, wear tight-fitting clothes or boots—as tight clothing hampers the circulation—and are out in the cold and fatigued from walking around or doing other activities because their body expends energy doing those things rather than trying to maintain the proper internal temperature.

Also, certain diseases (like diabetes) and medications, alcohol consumption, and smoking increase the risk for frostbite (Skinsight.com n.d.-a).

If you suspect a case of frostbite:

- Bring the patient indoors immediately.
- Do not thaw, apply direct heat to, or rub the affected part. Dip the affected part in warm water for about 30 minutes and treat it gently. Do not break any blisters if present.
- Place cotton balls between fingers and toes after warming.
- Wrap the area with a clean bandage loosely so that it does not freeze again, because the wrap should promote more heat for the area.
- Painkillers like ibuprofen and paracetamol can be taken for pain relief as directed (Kidshealth.org 2018).

Drowning

Drowning can occur when the nose and mouth become submerged under water such that a person cannot breathe. It can happen in the middle of a lake due to fatigue when swimming, or as an accident. When a person drowns, water enters the lungs. Due to this, the lungs become heavy. After 2 minutes, the patient loses consciousness. During the initial phase when the person becomes unconscious, they can still be revived by CPR and breathing assistance (see pg. 116). This stage only lasts for a few minutes. However, if intervention is not carried out in this stage, it can progress to hypoxic convulsion, i.e., the body shows abnormal body movements like jerking due to the decrease in oxygen content in the brain. This entire process takes about 10 to 12 minutes before death ensues.

The following measures can be taken if a person is drowning:

- Call 911 or any lifeguard if you see one nearby.
- Obtain consent if possible. However, it likely will not be possible.
- Take the person out of the water.
- Put on PPE if possible.
- Assess for neck injuries.
- Look for signs of breathing.
- If the person is breathing, check for signs of circulation (see pg. 52).
- If there are no signs of breathing or circulation, start CPR immediately (see pg. 116).
- Repeat the procedure if the person is still not breathing.
- If the patient starts vomiting, roll them onto their side to clear the airway.
- Treat the patient for hypothermia after they are breathing and have a pulse (WebMD 2021).

Takeaway

This chapter dealt with the health conditions that can occur due to difficult environmental conditions. While some of the conditions can present with mild symptoms, some other situations are life threatening emergencies. Hence, it is necessary to recognize these situations and manage them accordingly. Also, to prevent these occurrences, it is necessary to know how to find shelter in extreme weather. This is discussed in the upcoming chapter. Make sure to go through it so that you are fully prepared for environmental challenges. We wish you happy learning!

10
WHERE TO FIND SHELTER IN EXTREME WEATHER

In the previous chapter, we went through all the health hazards that could occur in the wilderness due to extreme environmental conditions. In this chapter, we will discuss how to protect yourself from the harsh environment, i.e., how to build a shelter.

Once, we went out on an expedition to search for wild bees. We had to go deep into the forest looking into every tree hole and canopy. Suddenly, the sky turned dark. The wind was howling at us and soon there would be a thunderstorm. We had no time to turn around and get out of the forest, and by the time the thunderstorm stopped, it would be dark. Among our group was a fellow author who had worked on a similar book to ours. Thanks to the information he gained while working on the book, he started making a shelter of his own. The shelter protected us the entire night from wild animals and the bad weather. It gave us a new experience of camping out in the wild. Moreover, it added fun to our trip and gave us a morale boost for surviving in the dense forest.

The skill that our friend had helped us save our lives. It is these skills that we will be talking about in this chapter. This can be practiced in your backyard so that your entire family can be prepared in advance.

Shelter is one of the necessities of life, along with food and water. It keeps you safe from extreme weather conditions, wild animals, and insect bites. When you are out there traveling in the woods, things can get rough for you. This is when a shelter can mean the difference between surviving and not surviving. It

Where to Find Shelter in Extreme Weather

not only safeguards you but also helps you enjoy your trip better and provides motivation for surviving the harsh conditions.

You may be thinking that you are a strong adult who can survive a few days of bad weather. But things may not be as straightforward as you think. Also, what can be easy to handle for you may not be the same for your younger or older family member (Telsonsurvival.com n.d.).

Before we begin with the actual process of making a shelter, let us first start by learning to make knots. Knots help us fasten the components of the shelter together into place. Some of the most commonly used knots with their illustrations are next (Fouche 2022; Krebs 2017):

Overhand Knot (aka Thumb Knot): This is the simplest and the most basic of the knots. We form a loop and insert one end of the rope through the loop. Then, the ends are pulled to secure it. It is the primary knot that acts as a stopper and jams an object into place.

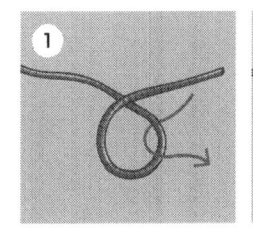

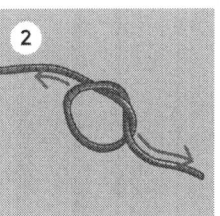

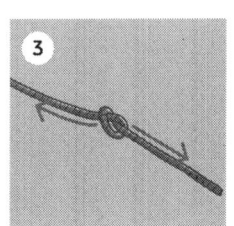

Overhand Knot

Double Overhand Knot: This knot is a modified version of the overhand knot. We make the loop twice, as the name states. It helps in securing the object even more.

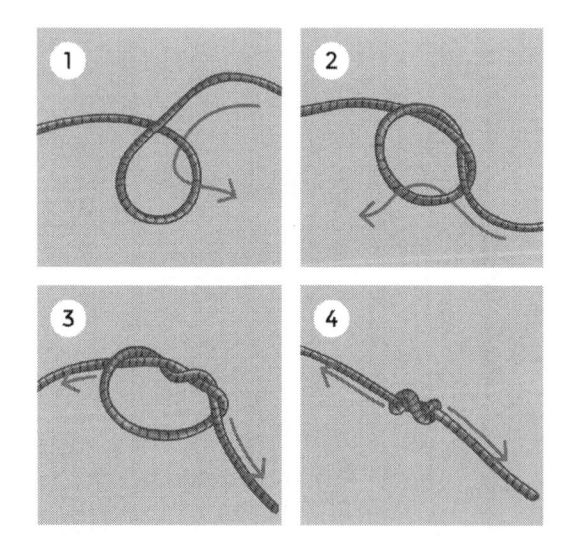

Double Overhand Knot

Reef Knot (aka Square Knot): The name "reef knot" comes from the use of this knot by sailors. This knot is used when we have to tie two segments of a rope together. We can tie this knot when we have to secure something which is not expected to move around often. However, the locking of this knot is not very strong. So, it is better not to use this knot to tie critical objects.

Reef or Square Knot

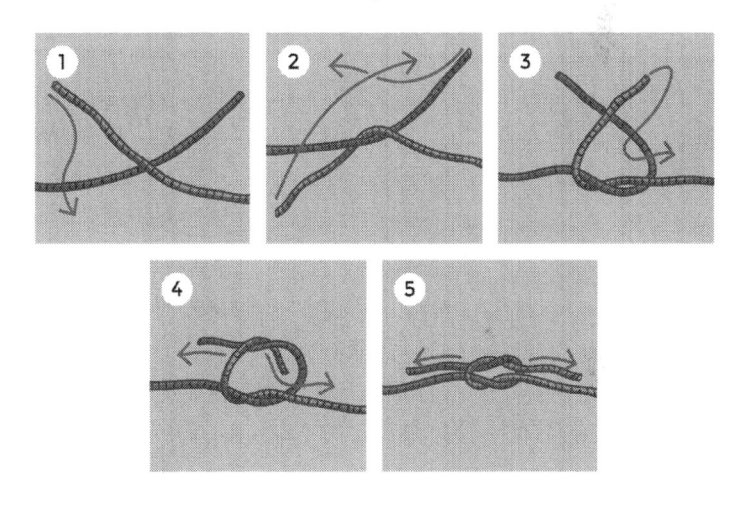

Figure 8 Knot (aka Flemish Knot): This is quite similar to the overhand knot (see pg. 223) but much easier to untie. When weight is put on this knot, it tightens up and becomes more secure. As the name suggests, start by making a loop. Continue over and around the standing part of the rope. The standing part is the section of rope which is horizontal and leads to the loop. Pass the end through the loop and pull the ends to secure (Fouche 2022).

Figure 8 or Flemish Knot

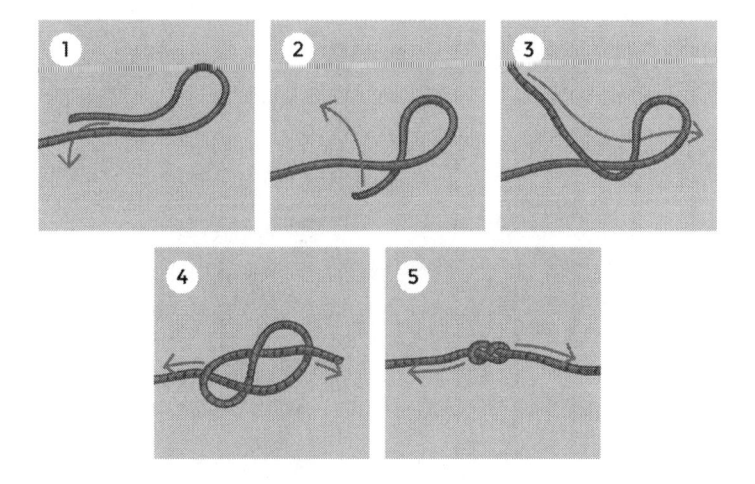

Figure 8 on a Bight (aka Figure 8 Loop or Flemish Loop): "Bight" in knot terminology means a U-shaped section of the rope. This knot need not be created with the ends of the rope like in the previous ones. It is an effective knot to secure objects and can be extremely difficult to untie. To make this knot, make a bight end of the rope. Then, pass it under itself to create a loop. Continue around the standing part of the rope and pass the end through the loop. Secure the knot by pulling on both ends (Fouche 2022).

Figure 8 on a Bight Knot

Figure 8 Follow: This knot is the most widely used in rock climbing, camping, firefighting, and loading heavy objects, as it is extremely strong. This knot is tied directly over an object, and due to its strong grip, it may be very difficult to loosen. First, tie a loose figure 8 knot by leaving a significant length of the extra tail. The knot can loosen its grip slightly in the initial stage of loading. The tail is then looped around an object and again, a figure 8 knot is made, but in reverse. After that, take the end behind the large loop and exit alongside the standing end of the rope. Pull on both ends to secure it (Fouche 2022).

Figure 8 Follow Through Knot

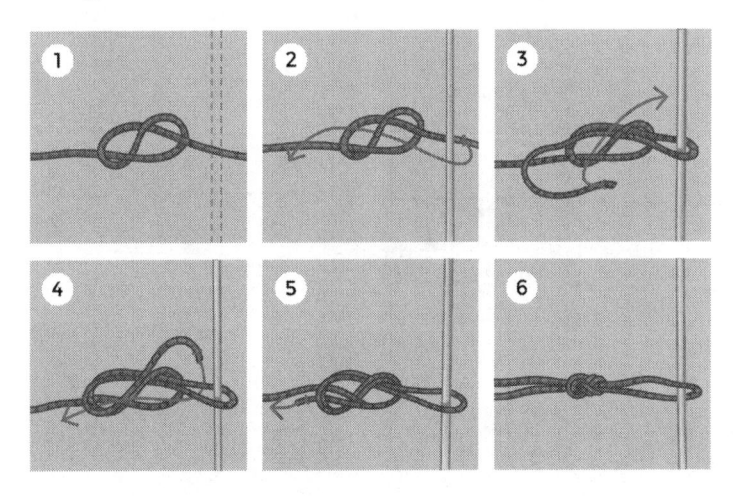

Bowline Knot: This is a reliable, secure knot that is easy to tie and untie. The main advantage is that it can be tied and untied with one hand only. This feature is very useful in wilderness emergencies if a person sustains an injury to the arm. However, as it can easily be untied, for security, it is better to use another reliable knot like an overhand knot after the bowline knot. To make this knot, make a loop with an end of the rope. That end is passed through the loop, around the standing part of the rope, and back through the loop. Pull on both ends to tighten it.

Bowline Knot

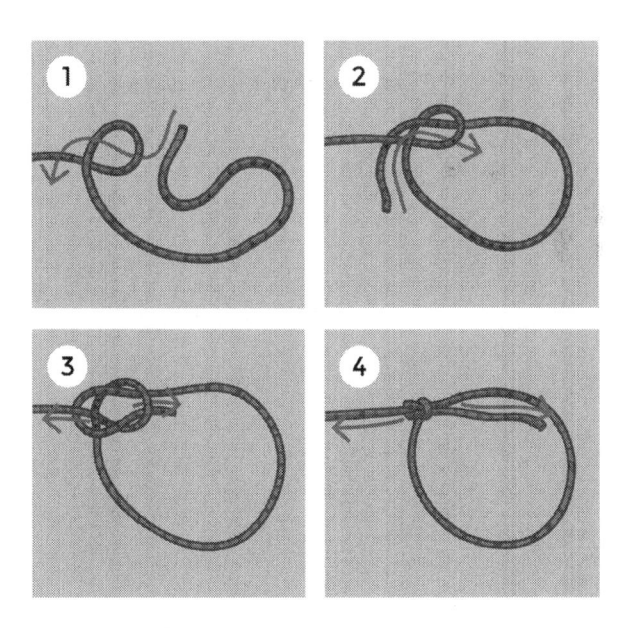

Round Turn and Two Half Hitches: This knot is tied around a fixed object and is extremely secure. It prevents the rope from sliding through or along with the object. It is easy to tie and untie and is used in boating, tying a rope to make a swing, or making a flybridge. To tie this knot, pass the rope around the fixed object two times. After that, cross the end over the standing part of the rope and pass through the loop. Repeat this procedure to form two hitches. Pull on the end and tighten it (Fouche 2022).

Round Turn and Two Half Hitches

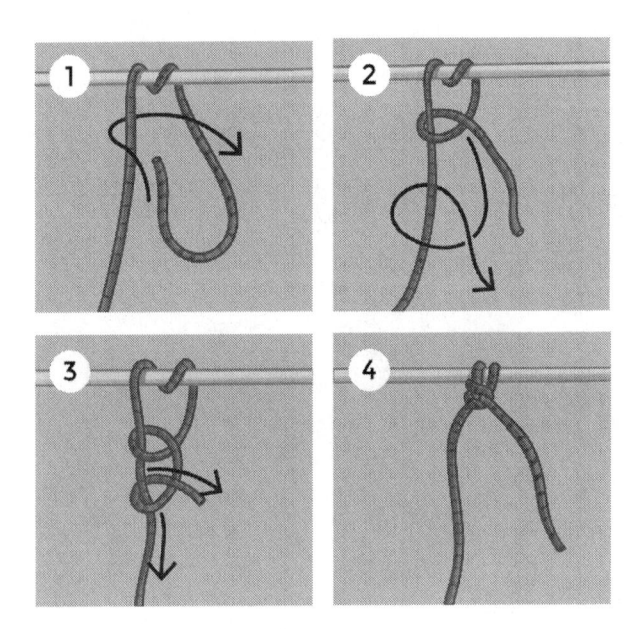

Halter Hitch: This knot is very easy to tie and untie. In this, we pass a rope around a fixed object like a pole. Hold onto one end and make a loop by going around the other part of the rope. Through the loop, another loop is made from the standing part of the rope and passed through the previous loop. This gives you your knot. We can easily tighten the grip on the pole by pulling on the longer end.

Halter Hitch Knot

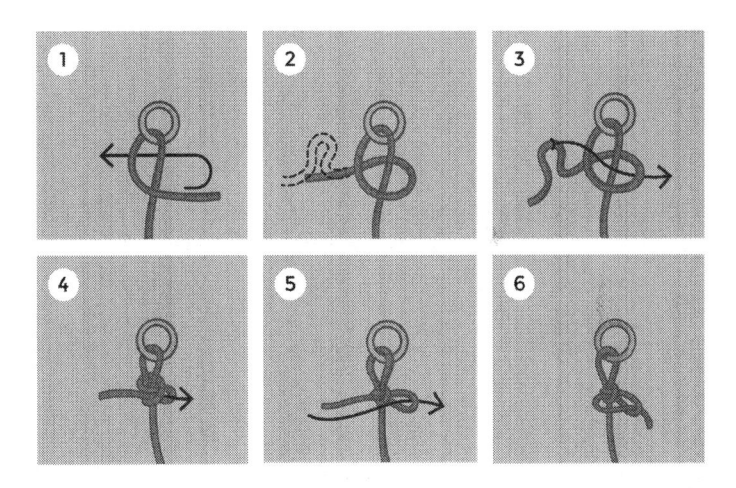

Double Sheet Bend: This knot is made when we have to join two ropes that are of different diameters. The knot prevents slipping and also helps to secure it in place. To make this knot, insert the thin-diameter rope through a loop made by the thick-diameter rope. Wrap the thin rope around the curved part of the thick rope and tuck it around itself. Repeat the same step to make a double knot. You can tighten the knot by holding onto the thick end and pulling on the thin ends of the rope.

Double Sheet Bend Knot

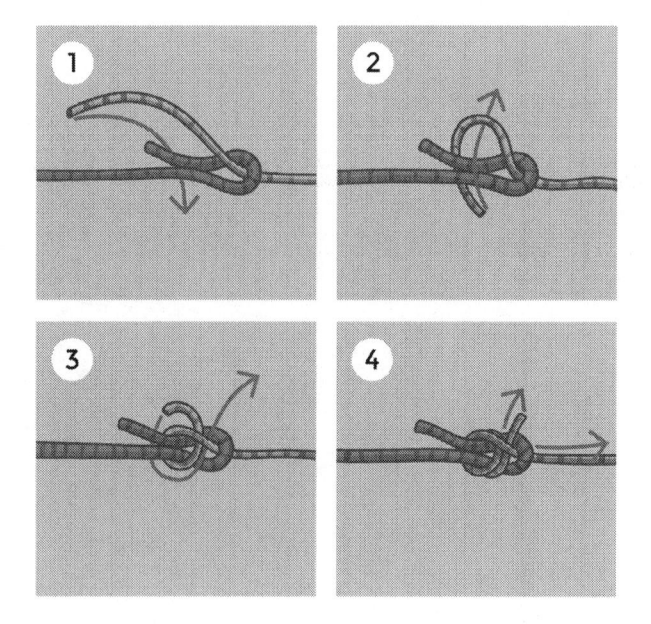

Clove Hitch: The clove hitch knot comes in handy when you need to tie lashings which are needed to build shelters (see pg. 241) This knot can easily untie and slip, so it should not be used alone. However, it can be used to adjust the height of the cloth attached to it, like that of a curtain, as the height of the bar can be adjusted by rolling the knot slightly. To make this knot, pass the end of the rope around a pole or fixed object. Do this for a second time. Then, thread the end under itself and pull on the ends tightly to secure it (Animatedknots.com n.d.-a).

Clove Hitch Knot

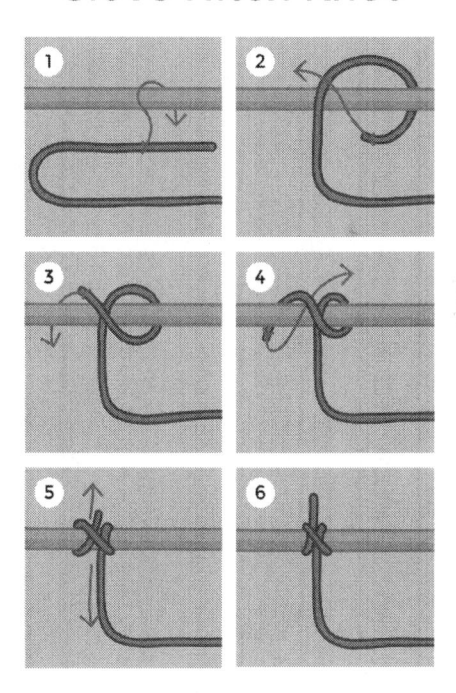

Timber Hitch (Bowyer's Knot): This knot is also known as the bowyer's knot, as it is used to tie the ends of the string on the ends of a bow; it's also used in certain instruments like the guitar. The major advantage of this knot is that it is used to transport cargo, as it is very secure when transporting loads but disassembles itself when loosened. To make this knot, pass an end of a rope around the object you want to tie and then around the standing end. Repeat this process three times and tighten it (Animatedknots.com n.d.-e).

Timber Hitch Knot

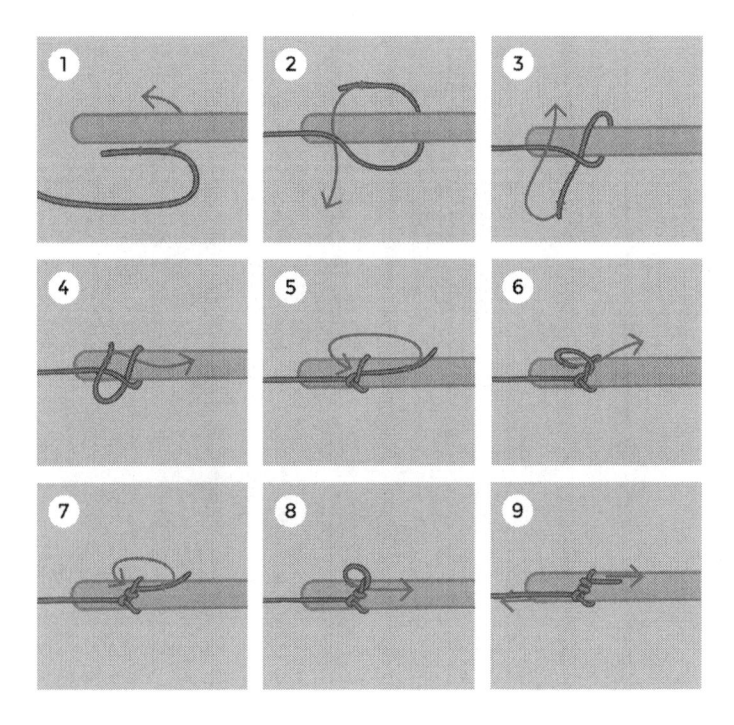

Lashings

Before starting with the lashing knots, what does "lashing" mean? Lashing means tying two objects with a rope to bring them together (Thefreedictionary.com n.d.). In this section, we will see the usage of the other two terms: "wrap" and "frap." Wrap means to tie the rope around objects, whereas the technique of passing and winding the rope around itself, in between the poles, is known as frapping. Frapping prevents the poles from rotating as well as helps in tightening the knot in place (Hutchison 2016).

Now, let us learn how to tie the three most common lashings.

Square Lashing

The square lashing knot is made when two poles are at a 90-degree angle. However, you can also use it in case of lesser degrees down to 45. Hence, the square lashing knot is used in tying a fence, making a raft, shelter, etc. (Animatedknots.com n.d.-d)

A clove hitch knot (see pg. 233) is tied around a pole. Tie this knot below where the poles meet at a right angle. Then, start with your first wrapping as shown in the illustration. Continue until three wraps are completed, then begin with your first frapping. Make three frappings and then finish with another clove hitch knot.

Square Lashing Knot

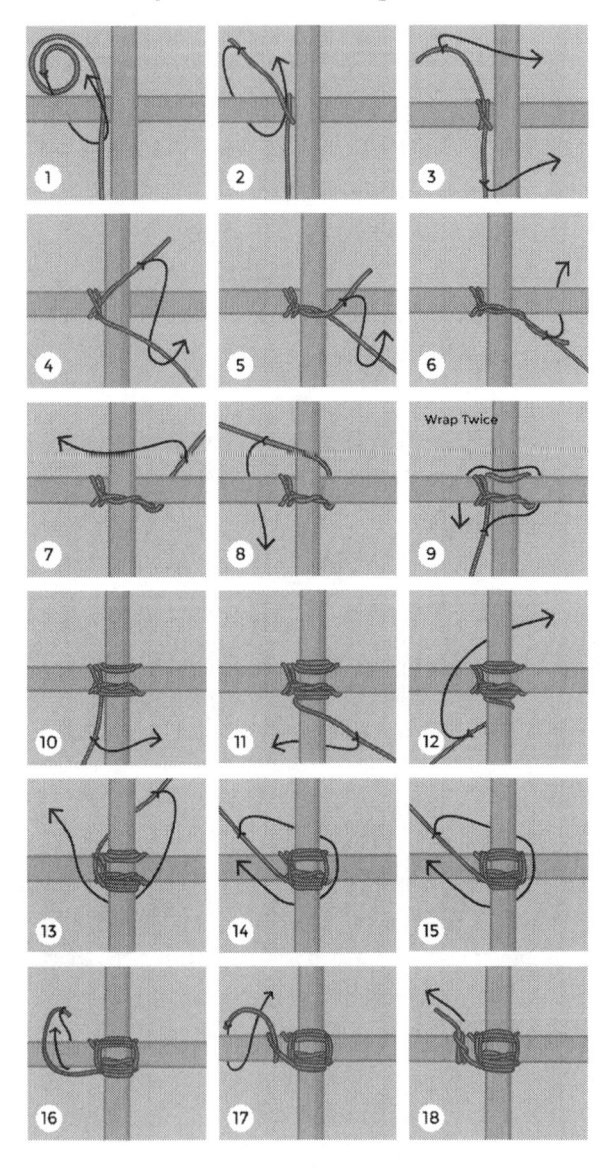

Diagonal Lashing

As the name suggests, this knot is made to secure two diagonal poles which cross at a variety of angles. It prevents twisting or rotation of the two poles. To tie this knot, start with a timber hitch knot (see pg. 234) on the top of the crossing. Next, start with your first wrap and continue with three more wrappings. Then, start with your first frap and make three more frappings between the poles to secure the knot in place. End your knot with a clove hitch knot (see pg. 233).

This knot is used in securing the diagonal poles when making a rectangular frame (Animatedknots.com n.d.-b).

Family Survival Medicine Handbook

Diagonal Lashing Knot

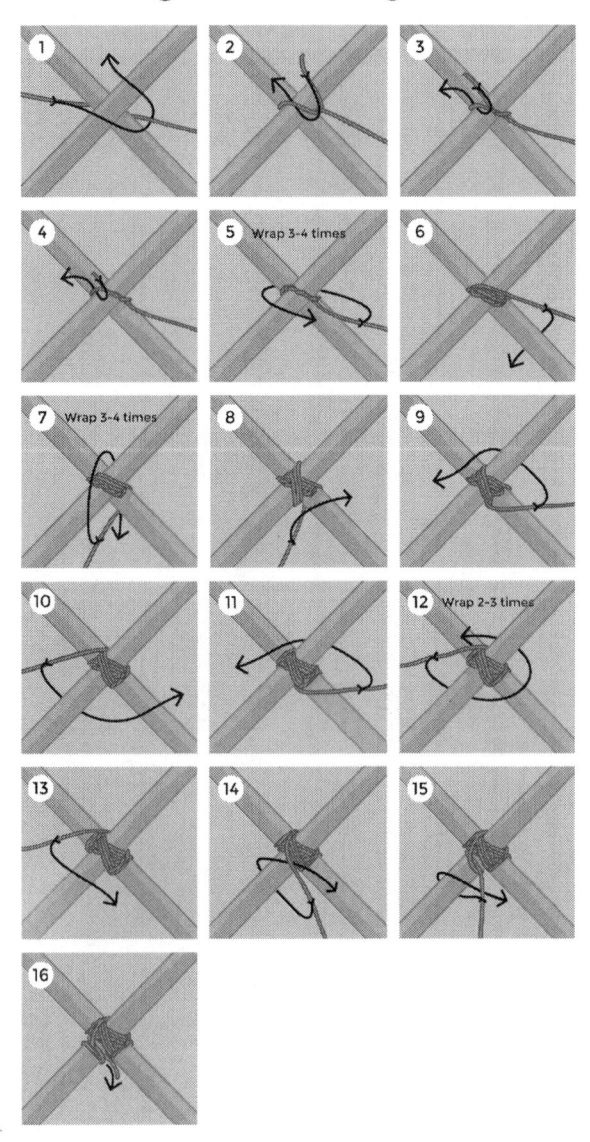

Shear Lashing

This knot is tied when binding two poles together such that the poles lie side by side. To make this knot, start with a clove hitch knot. Then, make six wraps followed by two fraps. Make another clove hitch knot and complete your lashing. You can then widely separate the legs of the poles to make the "A" shape (Animatedknots.com n.d.-c).

Shear Lashing

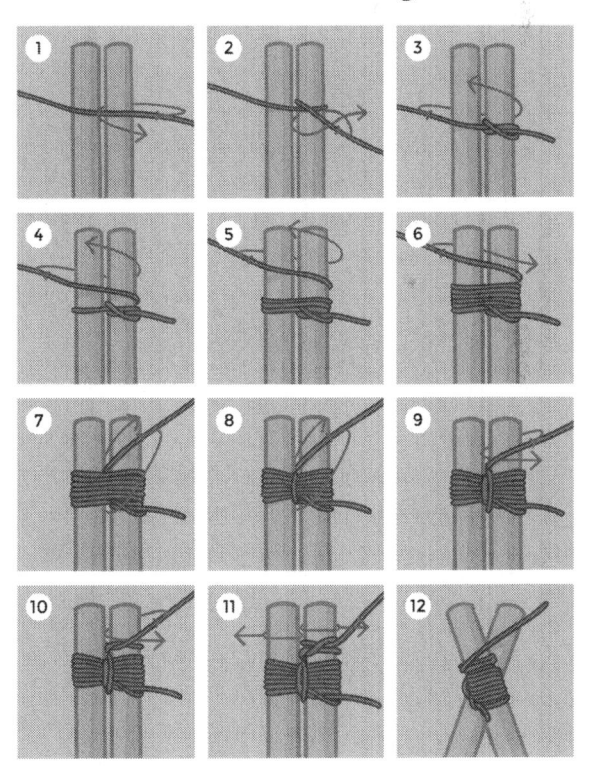

Now that we have learned the basics of the different types of knots and lashings, let us move on to their usage in making shelters.

What are the things to consider when building a shelter?

Where should you build a shelter?

You will want to build a shelter on dry, flat land with no water around. If you are hiding from something or someone, you will want to choose a spot somewhere hidden within the natural foliage. Do not build a shelter in a damp area, as it will be harder to stay warm or make a fire, and also building a shelter in a damp area will be more difficult than on dry land. Choose flat land, as you do not want water pooling or a place with a risk of landslide. Although nearby water resources may seem tempting, try avoiding this as there may be a risk of floods.

For how many people should a shelter be built?

One shelter is good for 2-3 people if you want to stay cozy without being overcrowded (Telsonsurvival.com n.d.). A large shelter is very difficult to keep warm. Hence, build a few more shelters if there are many of you. If you are alone, try making your shelter comfortable and small enough for you.

What is the situation?

You may want to analyze the situation and build shelter accordingly. For example, you have to analyze if you require the shelter for temporary or permanent residence and if you want it to protect you from harsh weather conditions or wild animals or both. If it is extremely windy, you may need a wind barrier; if rainy, you may require natural roofing or dense roofing to prevent seepage. If the weather suddenly goes bad, you may not have enough time to build a shelter. So, you will have to find a natural one. Do a quick survey of all these factors and plan accordingly.

What resources do you have?

Another important factor to consider is the availability of resources in the environment. You have to use and improvise with the natural resources to build a shelter to the best of your ability. We will be discussing the types of shelters and the resources that can be used in the later parts of this chapter.

10 Types of Shelters

We will be taking you through a step-by-step approach to making different types of shelters. The knots and lashings we discussed will help you with this section.

Lean-To Shelters

As the name suggests, the materials to build this shelter lean onto a support like a tree or a pole. There are two types: single lean-to and double lean-to shelters. In the single type, only one side of the shelter is built in a lean-to fashion. In a double type, both sides are made that way. Although a single lean-to shelter can protect you from weather conditions, it is less reliable and also cannot keep you warm. Here's how you can make a lean-to shelter:

You will need a pole or a wood log long enough to be kept horizontally between two trees. You can adjust it in the Y-shaped joint of the trees. To make this strong, you can use square lashings with a strong rope to tie it on both ends to the trees. Now, cover the sides (one side for single lean-to and both sides for double lean-to) with poles or wooden logs and branches diagonally, so that they meet the horizontal pole or log. You can then use leaves and foliage, blankets, or a tarp to cover the logs so that your shelter remains cozy and warm (Outdooradventureguide.co.uk 2017).

Example of How to Build a Lean-To Shelter

4 feet
1.21 meters
floor

1. Find the right place to set up shelter.

2. Find a tree limb that is easy to adjust and secure it to your chosen tree using a square lashing or double knot.

7 feet
2.13 meters

3. Obtain more tree limbs and place them at about a 45-degree angle to create the structure of your shelter. Place a log in front of them on the ground so they are held in place.

4. Continue to add smaller branches to your shelter frame to close all gaps.

5. Further close your shelter by covering the bottom 12 inches with leaves, moss, more branches, etc.

6. Clear the ground you will sleep on and use something like leaves, moss, and branches to create a barrier between you and the ground.

Tarp Shelter

Tarps are water- and fire-resistant sheets made of polyester. These are largely used for the making of temporary shelters. Simple tarp shelters can be made in a similar process to that of lean-to shelters (see pg. 242), i.e., we place a pole between two trees and cover it with a tarp. We secure the edges quite far from the middle so that we have enough space inside. However, this shelter cannot be kept warm, so its "modified" version is required. This modified version is known as a "tarp tepee shelter."

To make a tarp tepee shelter (also known as the parachute tepee shelter), first, take three long poles and tie them with a rope at one end. Use this rope to pull the frame diagonally until the entire structure is steady in place. Then, gradually add poles to the structure one by one and subsequently tie them with a strong rope. Now, take your tarp, raise it to the height of the frame, and wrap it around the poles. If the tarp does not completely cover the roof, you can use another tarp to cover it, forming a "parachute canopy."

This tarp tepee shelter will have enough space inside so that you can make a fire to keep you warm. But make sure you have enough ventilation at the top if you intend to build a fire (Bhaddock 2012).

Where to Find Shelter in Extreme Weather

Example of How to Build a Parachute Tepee Shelter

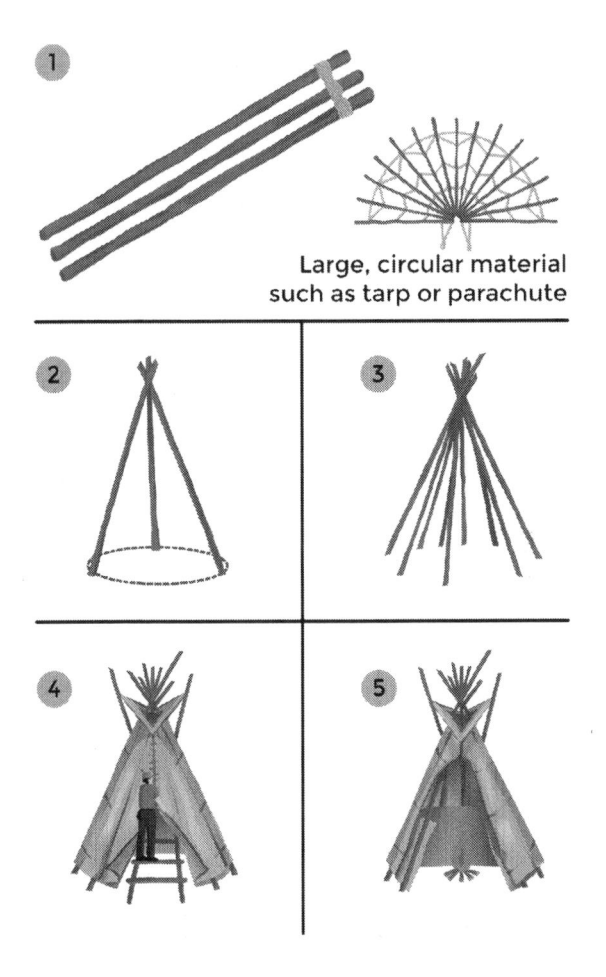

1.

Large, circular material such as tarp or parachute

2.

3.

4.

5.

245

A-Frame Shelter

The A-frame shelter is made up of ridges and poles that should be long and strong enough to support the entire frame. It is somewhat similar to the double lean-to shelter but more rigid and strong.

First, take two poles and tie them diagonally in the shape of an "A." Use timber hitch lashings with a few frapping turns, and then finish with a clove hitch. Repeat so that you have a second set of poles shaped like an "A." Then, take another pole and place it horizontally above the junction of the two poles. Secure it with the other two poles.

Take another pole approximately midway down and secure it to the logs, parallel to the horizontal pole, using a clove hitch knot and a few frapping turns (square lashing).

Take foliage, leaves, and whatever is available in your surroundings and weave along the sides so that you have a shelter that will keep you warm (Finchamp 2019).

How to Build an A-Frame Shelter Example

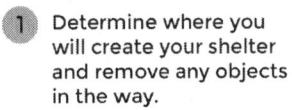

1 Determine where you will create your shelter and remove any objects in the way.

2 Obtain 7-9 strong branches that are 7-8 feet long. Ensure at least 5 of them are sturdy enough to hold up your shelter.

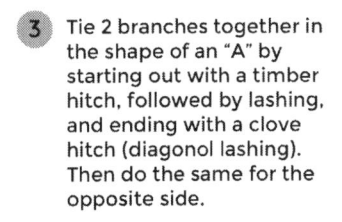

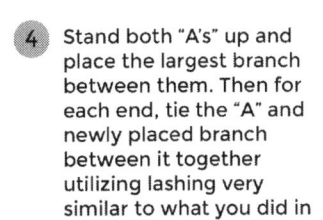

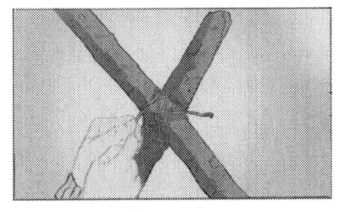

3 Tie 2 branches together in the shape of an "A" by starting out with a timber hitch, followed by lashing, and ending with a clove hitch (diagonol lashing). Then do the same for the opposite side.

4 Stand both "A's" up and place the largest branch between them. Then for each end, tie the "A" and newly placed branch between it together utilizing lashing very similar to what you did in the previous step. Make sure the paracord (type of rope) is wrapped around each leg and over the newly placed branch between them.

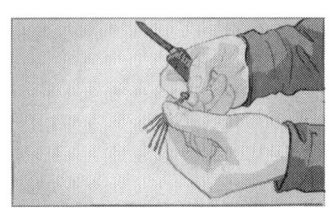

Family Survival Medicine Handbook

5. Create holes in the ground to insert each leg of your shelter. This helps it not move especially while still building it.

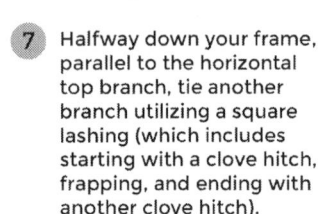

6. If limited paracord (type of rope) is available, cut it to have more.

7. Halfway down your frame, parallel to the horizontal top branch, tie another branch utilizing a square lashing (which includes starting with a clove hitch, frapping, and ending with another clove hitch).

8. Replicate this on the other side of your shelter and if more firmness is desired, add another branch of this nature on each side for a total of 4.

9. Collect branches with as many leaves and needles as possible.

10. Intertwine collected branches, leaves, and needles into your frame.

Snow Trench

A snow trench is a shelter made in the snow. It is the simplest and easiest shelter to make in the snow. A long, narrow ditch (trench) is made in the snow to a length that is a little more than your height. It should be about 3 feet deep. One snow trench is made only for one person.

Then cover the length of the trench with logs of wood or poles along the width. Place wooden sticks and branches along the length above the logs. You can then place a tarp above this frame and further top it with snow to test its strength and also secure the tarp. The snow trench will have an opening for you to enter. On the floor, you can insulate your trench by placing a sleeping pad, a mat, or leaves.

After you have entered the trench, you can cover the opening with a plastic bag full of snow. However, make sure to ventilate (see pg. 252) your trench (McCann 2011).

Snow Trench

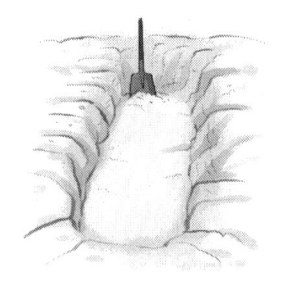

Create a trench that is about 3 feet deep and a little longer than you. In the event snow is not thick or deep enough, use the snow you dig up to build up the sides.

Cover the width of the trench with branches and sticks before adding perpendicular branches on top of that to cover the length of the trench.
Note: If you do not have a tarp, the placement of the branches and sticks for this step must be closer together to keep the snow you will add on top later from leaking into your shelter.

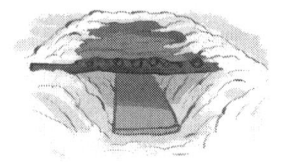

If available, place a tarp on top of the branches and sticks. Use excess snow to hold it down around the edges.

First, place snow on top of the tarp, if available. If a tarp is not available, place snow on top of tightly packed branches and sticks. Then place something on the ground to keep you warm. This could be an inflatable mattress, branches from trees, or a bag filled with debris.

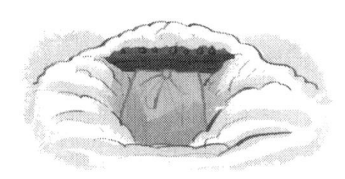

Block the entrance of your trench with a bag filled with snow, for example.

Snow Cave Shelter

A snow cave is a T-shaped cave dug into the snow. It will protect you from the worst snowstorm. To dig this snow cave, you will need a snow height of at least 3 feet. Start by making an entrance as shown in the illustration on pg. 253. The entrance should be 36 inches in height and 18 inches in width. Next, widen the top of the entrance into a T-shape. This widening should occupy at least 18 inches of the total height of the entrance and should be 66 inches in width (see illustration pg. 253). Then, dig further into the drift, at least the length of your body, to make sleeping areas on the two sides. After you have done this, seal the horizontal parts of the "T" with snow blocks (Scoutlife.org n.d.).

The cave is not done unless you have ventilation. Ventilation is the exchange of indoor air with outdoor air. The indoor air contains breathed-out carbon dioxide and several other indoor pollutants. Ventilation acts to dilute the indoor air and also helps in the circulation of gases (Lung.org n.d.). To maintain this ventilation in a snow cave, poke multiple vent holes at approximately 45 degrees to the floor. You can use a ski pole or shovel handle for this purpose. Then, you can keep a bag filled with snow at the entrance. It will keep you safe (Scoutlife.org n.d.).

How to Build a Snow Cave Shelter

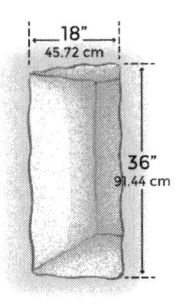

Find a large hill of snow that was wind built (snow drift) or part of a hill or mountain that is stable and covered by snow (snow slope) and dig out an entrance that is at least 18" (45.72 cm) wide and as tall as your chest (in the picture noted as 36")

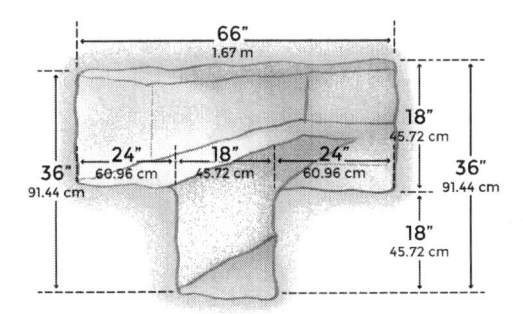

Around waist level (in the illustration 18"), increase the width of the entrance by cutting or digging out 24"(60.96cm) to the right and 24" to the left (60.96cm) of the entrance. Once complete, your entrance now looks like the shape of a "T."

253

Family Survival Medicine Handbook

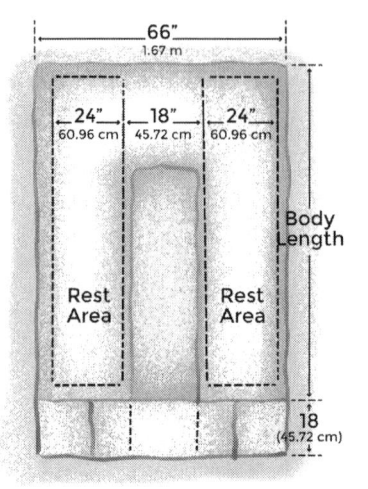

Diagram labels (aerial view): 66" / 1.67 m; 24" / 60.96 cm; 18" / 45.72 cm; 24" / 60.96 cm; Body Length; Rest Area; Rest Area; 18 (45.72 cm)

Aerial view: Behind the top or horizontal portion of the "T" you just created, dig out a dome shape large enough for two people, one on either side, to lie down (length) and sit up (height). The floor of your cave should be around waist height thus a lot of your digging will occur while on your knees.

Cover the horizontal portion of the "T" entrance with snow blocks, snowballs, or bags of snow.

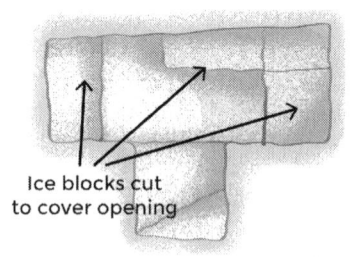

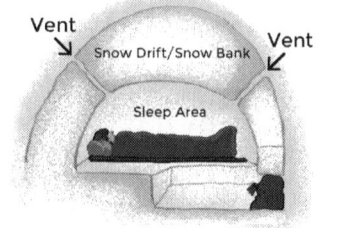

Ice blocks cut to cover opening

Diagram labels (side view): Vent; Snow Drift/Snow Bank; Vent; Sleep Area

Use a pointed object such as ski poles or shovel grip to create proper ventilation at a 45 degree angle from the floor. And use something like a bag of snow to cover the open vertical portion of your entrance.

254

Igloo

An igloo may seem quite difficult to make. But it will also provide long-term shelter. You will need a few preparations beforehand. Pack a plastic bin or any container with snow. Then, while maintaining form, release the contents. The snow blocks can be about 3 feet (1 m) long, 15 inches (40 cm) in height, and 8 inches (20 cm) deep. The base of the igloo needs to be approximately 2.5 feet deep. For the floor, about 6 inches of snow can be left as it is (VertDude n.d.).

After snow bricks have been readied, start piling them to form a circle. Try to make the circle an equal distance from the center on all sides. Build up walls in such a way that they lean inward. Stack the bricks until you reach the top, where you need to make a roof with ice bricks. To make a roof, make a block slightly larger than the hole. Place the block on the top, then, from the inside, wiggle it so that it fits exactly in the hole (Reddit.com 2018). Make a cold sink under the wall so that the cold air can fall and flow to the outside (Raveendran 2022). Put several blocks along one wall as a sleeping platform. Make ventilation (see pg. 252) ports in multiple places on the roof with an ice axe. At the entrance, make a tunnel and put a roof above it. This will prevent snow from blowing into the igloo. You can seal the crevices formed between the ice bricks with more ice (VertDude n.d.).

Now, you have an igloo that will keep you warm in the snow. The next illustration sums it all up.

How to Build an Igloo

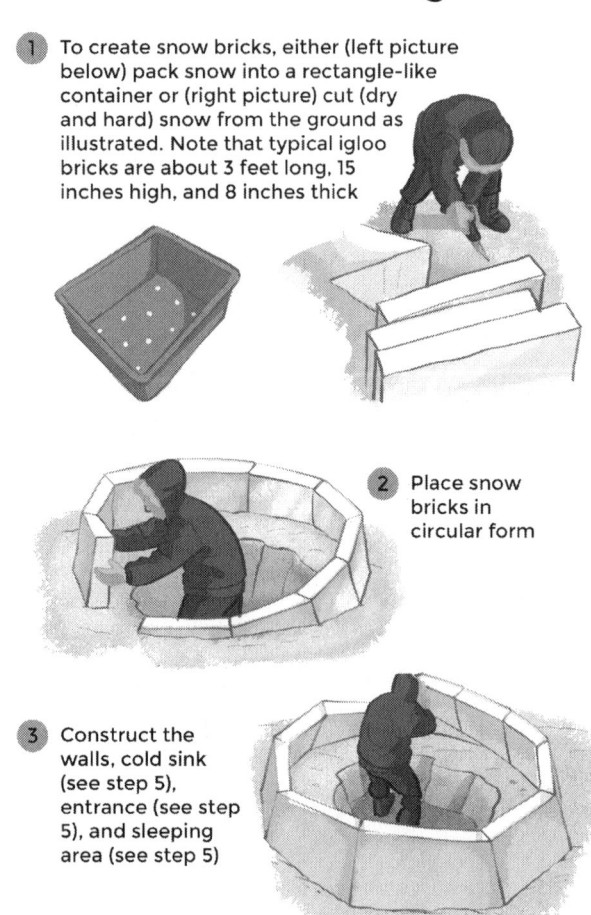

1. To create snow bricks, either (left picture below) pack snow into a rectangle-like container or (right picture) cut (dry and hard) snow from the ground as illustrated. Note that typical igloo bricks are about 3 feet long, 15 inches high, and 8 inches thick

2. Place snow bricks in circular form

3. Construct the walls, cold sink (see step 5), entrance (see step 5), and sleeping area (see step 5)

Where to Find Shelter in Extreme Weather

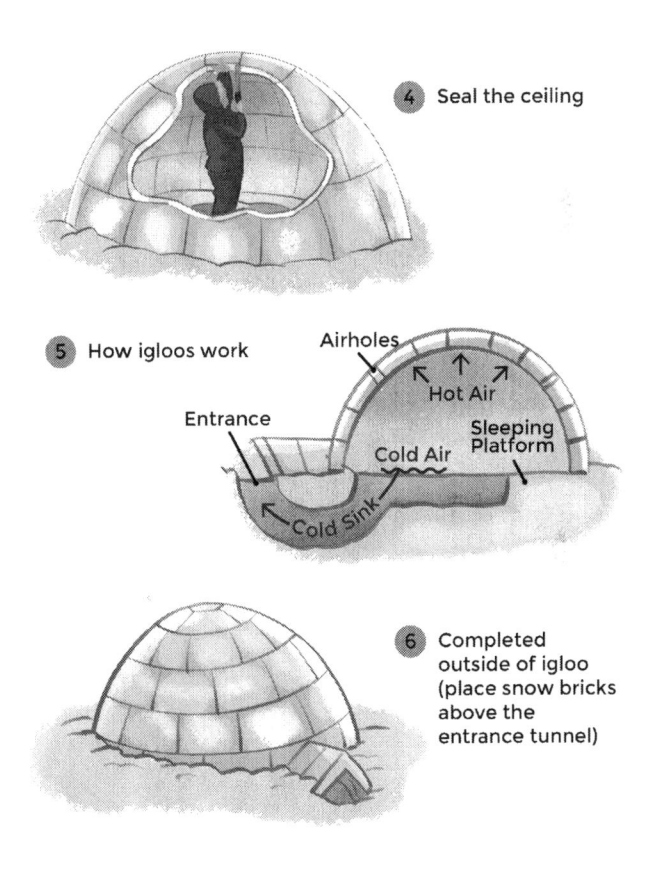

4 Seal the ceiling

5 How igloos work

Airholes
Hot Air
Entrance
Sleeping Platform
Cold Air
Cold Sink

6 Completed outside of igloo (place snow bricks above the entrance tunnel)

257

Tree Pit Snow Shelter

This type of shelter can be made in the woods in snow. It should be made under a bushy tree so that the leaves and branches can be used as a roof. Dig a pit in the snow around the trunk of the tree. The pit should be at least as long as you are tall. If you cannot dig at least to your shoulder height, pile up the ground on either side. The foliage should also be kept as a carpet on the floor. You can also add leaves and branches overhead, but make sure to provide enough insulation (Action-Hub Reporters 2011).

Tree Pit Snow Shelter Example

Evergreen Boughs (the main branch of an Evergreen)

Stuffed Snow

Stuffed Snow

Evergreen Boughs (the main branch of an Evergreen)

Ground Floor

Debris Hut

A debris hut is a modified form of an "A-frame" shelter (see pg. 247), differing in the fact that the horizontal stick of the A-frame (also known as a ridgepole) is placed on the ground. Thus, it will have only one opening. It is not possible to make a fire inside, so the hut should be made small enough for one person, to trap your body heat. Also, make sure to insulate the floor of your hut with leaves and foliage to a thickness of at least 4 inches. You can also place a tarp over the A-frame and then cover the tarp with leaves to ensure adequate insulation. See the illustration for details (Gearpatrol.com 2014).

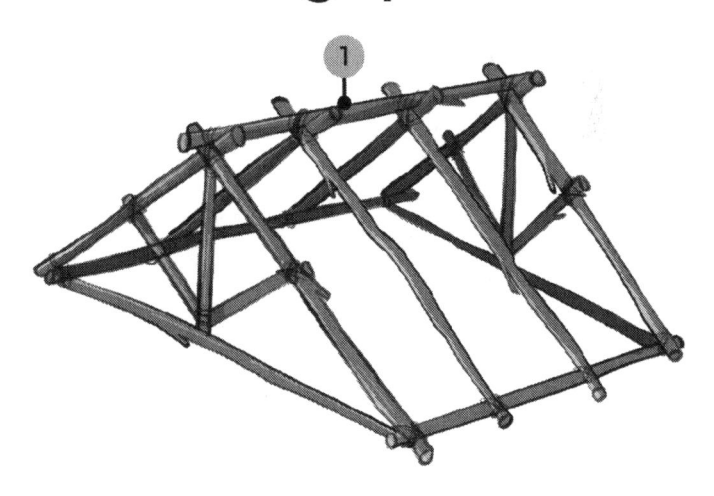

Debris Hut Example

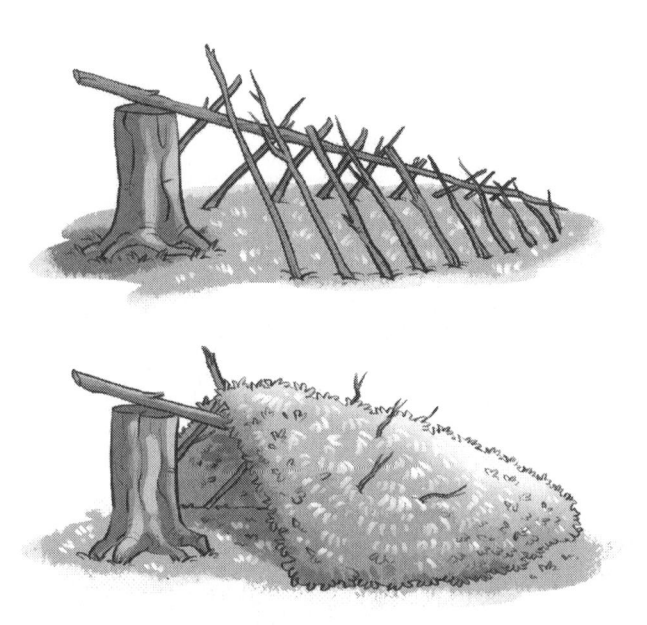

Dugout Instructions

A dugout is made as a shelter to protect yourself from the wild. You will have to dig a furrow (a long, narrow hollow) 8 to 10 feet deep. The entrance and exit should be sloping and the mud that you dig out while making the furrow should be kept aside. This will prevent anyone from knowing that you built a dugout. Make the available branches, leaves, and foliage your bedding so that it acts as insulation on which you can sleep. Use sandbags or branches to strengthen your wall. Cover the entrance with foliage in such a way that it remains camouflaged. However, keep the entry and exit open for easy escape (The Survival Journal n.d.).

Survival Strategy

The necessities of life, as we all know, are food, water, and shelter. However, in survival in the wilderness, a few extra points are added. For example, a fire is quite a necessity, and also mental preparedness is a survival strategy if stranded in the wild.

The following points should always be remembered as a part of survival advice (Michael 2021):

- Don't panic, and stay calm. We have already discussed the importance of mental preparedness in Chapters 1 and 2. Go through those sections again. And always remember that to perform the necessary actions when stranded in the wild, whether it be helping the injured, arranging for help signals, or making a temporary shel-

ter, mental calmness is necessary to make your actions quicker and more effective.
- Effective signaling can help rescuers identify you. There are various ways in which you can generate rescue signals. Create smoke by burning dry leaves and branches. Make large signs of "Help" in the sand, snow, or mud. Attach a cloth onto a stick and wave it in a clearing if you hear rescuers approach. Try reflecting light on a shiny object, which will help attract attention. You can also whistle if you have a whistle in your backpack.
- Treat injuries as soon as possible. If you sustain an injury, attending to it immediately can help the wound heal quicker and prevent future complications.
- Make sure you think about shelter, water, and food before you completely run out of them. You will learn how to purify natural water in Chapter 11.
- Fire is another very useful tool in the wild. However, make sure that it is made in a safe environment to prevent disasters. You can then use fire for multiple purposes like boiling water, keeping away wild animals, cooking food, and keeping yourself warm.
- Conserve energy. You will have limited resources in a scarce environment. So, make sure you conserve energy directly and indirectly. You can directly conserve your energy by avoiding exertion and talking less. Indirectly, you can prevent body stress by taking care of certain factors. In the cold, avoid shivering by staying in a warm environment. In the heat, avoid a lot of sweating by keeping yourself in a cool environment.

The less stress your body has to go through, the more energy you save.

Surviving Natural Disasters at Home

Throughout this chapter, we have been teaching you to build shelters when you encounter a harsh environment in the wilderness. However, what if you have a shelter? Is your home completely safe during a natural disaster?

Even though the home is relatively safe compared to the outdoors when you have a natural disaster, there are still possibilities of injuries. Hence, you will have to keep in mind the following factors if you want to survive a natural disaster:

- Always stay alert with updates on weather from your local authorities. Keep a battery-powered radio with you, if possible, so that you will have the updates even if the power goes off.
- If, before the disaster strikes, you have been informed by the local authorities to evacuate, do so immediately.
- Also, if you are in a temporary shelter or a treehouse, evacuate immediately.
- Keep stock of food supplies, water, and medical supplies beforehand. Also, have a whistle.
- Charge your cell phones to the fullest and only use them when necessary.
- After the disaster, if you are trapped or injured, use a whistle to notify the rescuers in the surrounding area.

- Stay away from fallen buildings and power lines, as these can be dangerous.
- Administer first aid or try to get medical help if you are injured (Uwc.211ct.org 2022; Ready.gov 2022; Smith 2017).

If there is a hurricane or a tornado when you are at home:

- Take shelter in the basement or the lowest area possible.
- Stay away from windows and keep the doors closed.
- Try to stay somewhere in the middle of the house, for example in the corridor/hallway, laundry room, closet, or a small, inner, confined room.
- Cover your nose and mouth with a cloth or handkerchief and avoid breathing dust, if present.
- Do not drive in this situation. Instead, stop the car, and cover yourself up with clothes or a blanket, if possible (Uwc.211ct.org 2022).

Takeaway

Until now, we have almost covered the most common strategies needed to survive in the wilderness or a disaster-stricken household if an emergency ensues. The strategies include maneuvers for examination, attending to mild, moderate, and severe injuries, and also strategies to tackle environmental emergencies by making shelters.

However, only reading this information once may not be effective when you have to use it. So, from today, make it a habit to practice this with your family. For example, you can practice how to build shelters every weekend, hold family quiz nights on the different types of shelters, and so on. With regular revision and practice, you will be prepared for most bad situations before they happen.

11
MEDICATIONS: THE ALTERNATIVES

Throughout this book, we have discussed the "unexpected" emergencies. You were not anticipating anything that was to come your way. You were outdoors where the environment was not familiar, or else in a household where, even though the environment was familiar, the expected and at times required resources were not available. We tried our very best to teach you how to approach and tackle commonly seen situations. We taught you how to be prepared for such events. However, the meaning of "being prepared" is not only limited to this. It also means being prepared when you don't have gauze, cotton, or medicines like ibuprofen and paracetamol with you. What are you supposed to do now?

In this chapter, we will try to provide more insight on the medical alternatives, i.e., creating alternatives for common materials used in the treatment of medical conditions. Let us go into these one by one.

Thus far, we have talked a lot about taking traditional antibiotics. Well, what if you are caught off guard and do not have your traditional prescribed or even over-the-counter medications? This section will offer you alternatives.

Here is a list of alternative antibiotics:

- Honey: If using on a wound, apply it directly. If using it for an internal infection, swallow a tablespoon or mix it into your herbal tea. If possible, use raw manuka honey. Note: it is not recommended for babies under 1 year old.

Medications: The Alternatives

- Garlic: 2 cloves per day is considered safe. Larger doses could cause internal bleeding and amplify blood-thinning medication. If rubbing on a wound, use your homemade garlic concentrate. It is made by soaking a few cloves in olive oil.
- Myrrh: Typically, it is prepackaged, so apply as stated to the skin. Overall, most people can use it but it is known to potentially cause diarrhea if ingested. Large doses could lead to heart problems.
- Diluted thyme essential oil: This can be applied to the skin. Equal parts thyme oil should be combined with equal parts coconut or olive oil. If applied without being diluted, irritation and inflammation may occur. It is not to be ingested by mouth.
- Diluted oregano essential oil: This can be directly applied to the skin. One drop of oregano oil to one teaspoon of olive or coconut oil is how to dilute it. It should not be ingested by mouth or used as undiluted oil on your skin (Brusie 2019).

Maybe you do not have ibuprofen, acetaminophen, or asprin or you want other natural alternatives. Look no further!

Here is a list of alternative painkillers:

- Ginger: It can be used as a topical cream, gel, or essential oil (make sure to dilute with olive oil or another carrier oil) or ingested via tea, capsule, or tincture. Make sure it is high quality and remember that

if ingested without food in a concentrated form, it can cause a stomachache.
- Turmeric: Research has found when taken by mouth or applied to the skin, less than 8 grams per day is safe.
- Capsaicin: It is available in gel or cream form and can be applied 3-4 times daily
- Valerian Root: Drink valerian tea to get nerve, thus pain, relief.
- Magnesium: It is available as a supplement or naturally through eating sunflower or pumpkin seeds. It has especially been found to help with migraines and muscle spasms.
- Cat's Claw: It is available in capsule form. 250-1,000 mg capsules 1-3 times per day is the recommended dosage, otherwise, it is known to cause diarrhea.
- Boswellia: Taken mainly as a supplement and used topically (Indian Country Today 2018; Cronkleton 2019; Newsnetwork.mayoclinic.org 2020).

Generally speaking, various studies have been conducted using varying amounts of these substances unless stated otherwise; thus, there is no guideline as to the exact amount that should be taken in a day. You can try taking one or two servings of any beverage per day and applying the ointments 2 times per day. Take these until you feel better and, of course, if you notice negative side effects, stop taking or applying them.

Water

The benefits of water are known to all. It quenches thirst, balances our body temperature, and without it, a person can only survive for 2-3 days. For medical purposes, we can, for example, use water to keep our wounds clean and stay hydrated. However, if we are out in the wild and cannot find water resources, what can we do? Here are a few tips that can help:

Rivers and water sources: The basic first step is to look for water sources. These may be located in the middle of a valley, as flowing water settles there. If legal where you live, you can always try collecting the rain. You can do this by funneling a flat surface into a container or by hanging out clothes that absorb the rainwater.

Plants: Try locating a source of water from plants. If you see patches of vegetation, try digging under them. You may find water underneath the layer of this vegetation. You can also try to extract water from the plants. In climates where there is heavy dew formation, this can be either licked or absorbed. You can tie plastic bags around the bushy leaves of a tree. The evaporated water droplets condense into the bag, which can act as a source of water.

Animals: Another way by which we can locate the source of water is by watching animals. Animals mostly graze where the water is near. Only some animals travel for a long distance to find water. Birds like pigeons are also found near water. If you see ants climbing up a tree, there may be a reservoir in the tree itself. However, the presence of reptiles and birds like eagles and vultures may not indicate a water resource nearby (Michael 2021).

Now that we have learned how to find and extract water from our surroundings, let us learn the measures we can adopt if we do not have the resources to boil the water.

Standard recommendations state that boiling water even for one minute kills almost all of the germs present in it (Health. ny.gov 2018). This is the best method by which we can purify water. However, in the wilderness, we may not find the resources required for boiling. Wood may not be available or may be wet so that we cannot burn it for fire. So, how do we purify water?

A natural source of water purification is sunlight. Sunlight carries ultraviolet rays, which kill the germs present in water. The water has to be kept in a transparent plastic bottle and kept in bright sunlight for 6 hours. If cloudy, it may need to be kept outside for 2 days.

Another method is the use of iodine tablets. These are easily available and easy to carry but may give the water the taste of iodine. Two iodine tablets purify one liter of water (Andrew n.d.).

All these methods can be adopted in the wild to purify water for drinking purposes.

Medicinal Plants Found in the Wild

Plants have been used for first-aid treatment dating back far into history. Some of the plants that can be used for medicinal purposes are:

Arnica	**Arnica** Arnica has small, yellow, daisy-like flowers. It can be used for topical application on the surface as an ointment but should not be consumed orally. It relieves muscle aches and bruises (Modernhippiehabits.com n.d.).
Burdock	**Burdock** Eating burdock is a great way to take in carbohydrates and the leaves have medicinal properties. You can steam the burdock leaves until they are soft and apply them to the affected area. This can treat bruises, bumps, and minor burns (Haigh 2019).

Calendula

Calendula
This herb contains yellow and orange flowers and is fragrant. Tea made with calendula is used for wound washing. It is also used to soothe burns, bee stings, rashes, and dry skin (Sarnacki 2019).

Cattails

Cattails
The stem and leaves of this plant can be crushed to extract a jelly-like substance. This jelly can be applied to soothe a burn. Also, the roots of this plant are filled with starch and are edible and nutritious. However, some have noted it left them with a stomachache (Sarnacki 2019).

Medications: The Alternatives

Common Plantain	**Common plantain** This plant is a weed found almost everywhere. It is used to heal cut injuries and wounds. Hot water extracts of this plant are also seen to be effective in boosting the immune system and are used to treat respiratory and gastrointestinal issues (Sarnacki 2019).
Common Yarrow	**Common yarrow** This is a plant with small white flowers and feathery leaves. The leaves of this plant can be crushed up and used to apply to wounds or burns. The tea made from its leaves can also be used to treat colds and headaches (Haigh 2019).

Dandelion

The dandelion plant has yellow flowers and is quite distinct. The flower is recognized as a great source of vitamins A, B, C, and D and can be eaten. It improves digestion and is an excellent source of potassium. When the roots are roasted, they are said to have a coffee-like look, therefore some people use it as a coffee. It may also be applied topically to treat cracked skin in oil form. To make the oil, harvest, clean, and dry dandelion flowers (air dry for 1-3 days near a window). Place them in a clean jar that is ¾ full of any oil (preferably olive). Stir the flowers to make sure they are all within the oil. Put a lid on the jar and place it on a windowsill. After 2 weeks have passed, remove the flowers from the oil or else they will start to mold. Now your oil is ready! If done properly, it lasts for up to a year. Note: if any mold or bad smell develops during the process, start over. Keep the prepared oil in a cool, dark place. (Haigh 2019; Adamant 2018; Adamant 2021).

Medications: The Alternatives

Goldenseal

Goldenseal
This plant is named after its yellow roots and sap. Extracts from this plant have been used to relieve upset stomachs and also as a topical treatment for skin and ear infections (Sarnacki 2019).

Jewelweed

Jewelweed
This is a tall plant with bright orange flowers. The sap from the stem and leaves acts as a remedy for skin irritation. You can crush the stem for several seconds, after which the sap starts oozing out. This can be applied directly to the skin (Sarnacki 2019).

Lavender

Lavender is an herb with fragrant, purple flowers. It is mainly used as a tea or taken in its extract (concentrated) form. It is used to relieve gastrointestinal problems and increase appetite. It also relieves anxiety and boosts mood. Lavender oil is known for wound-healing properties. However, some people are allergic to this plant and hence, it should be used with caution (Sarnacki 2019).

Old man's beard (Bearded lichen)

The long strands of bearded lichen have been found to possess antimicrobial properties (they protect from infection). These strands are either soaked in water and applied over the wounds or are dried and ground into powder form. This powder is then applied over the surface or can also be consumed by mouth (Sarnacki 2019).

Selfheal

Selfheal
Selfheal plants can be brewed into tea. It is seen to be effective against viral infections, the flu, mouth ulcers, and fever (Haigh 2019).

White Willow

White willow
The main part of the white willow tree which has medicinal effects is its bark. 1 to 2 teaspoons of the tree bark can be boiled in water for 20-30 minutes to make tea. Drinking this tea is seen to reduce inflammation (local reaction to injury and infection) and relieve pain. You can consume this tea twice a day until you start feeling better. In 1838, the extract from willow tree bark, salicin, was purified to form aspirin, which is still a predominant medication used for headaches today (Sarnacki 2019).

The medical data on the dosage and recommendations for the intake of these plants are not known, unless specifically stated. Just because they are natural does not mean you can consume them excessively.

Identifying Plants

In the previous section, we discussed the benefits of some medicinal plants. In order to identify these, the best place to look is in the woods. You must identify the plants with certainty before using them.

The books should be colorful with pictures and explanations so that you can easily compare the features and find the right plants.

Treating Flu and Fever

When in the wild, you can contract flu very easily. Flu is a viral infection in which a person can have a fever, headache, sore throat, runny nose, body ache, and fatigue. This can leave you feeling sick, tired, and fatigued. If anti-flu medicine is not available to you, you can opt for other natural measures. The treatment options are ginger and elderberry.

Ginger can be used as a tea or in a powdered form. It can be taken as an oral supplement or we can make tea from ground or powdered ginger and drink it. This helps relieve the cold and flu.

Elderberries work well as a flu treatment, but you need to use caution. To get rid of the bad/toxic material, you can pick and boil only the completely matured berries. The ripe berries, for instance, can be boiled to produce tea (Kps n.d.; Plantaddicts. com n.d.).

Elderberry

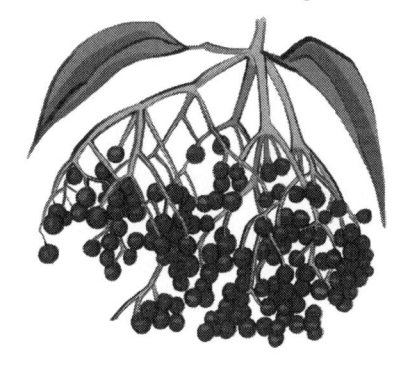

Various studies are being conducted as to the preparations, amounts, and frequencies to be taken per day but exact guidelines have not been determined. However, you can drink the beverages once or twice a day until you feel better.

Bandages From the Wild

If you sustain a cut injury or a bleeding wound in the wild, and you do not have any bandages or antiseptic at hand, what do you do?

Look in your surroundings and see if you can find seaweed. Seaweed contains iodine, which acts as an antiseptic and kills germs. These plants, when dried, become tightly wrapped around your wound and act as a bandage. They also stop bleeding. Baked, dried, and powdered seaweed are other forms in which the seaweed can be used. These forms, when applied to a cut injury, also stop bleeding, prevent infection, and protect the wound by acting as a bandage (Haigh 2019).

Seaweed

The next option is the use of a fungus of the birch tree, also known as the birch polypore. This fungus should be removed using a clean knife and cut according to the shape of the wound. The smooth side of the fungus should be stretched slightly and kept in contact with the wound. This will bind to itself and you can use it as a plaster (Hamilton 2009). A plaster is a strong dressing that prevents the movement of broken bones and joints until they heal.

Takeaway

This chapter is intended to give you some information about the alternatives available in the form of medicinal herbs or at-home remedies that can be used in wilderness emergencies. We hope it has provided you with useful, practical options to get you started on your survival prep journey or, if more experienced, keep you well informed.

CONCLUSION

That is the end of our presentation on Family Survival Medicine! We hope that you learned a lot and enjoyed! We want you to know that you can do this and use us as a resource for your foundation in survival medicine. We believe our content is so important because it could save a life and prevent a lot of pain. Thus, we ask again that you consider purchasing a copy of this book for your family and/or friends, and leave a review if you found any parts of this book helpful and worthwhile. As the old saying goes, a rising tide lifts all ships.

Thank you for allowing us to walk alongside you and your family on your journey. Please stay in touch by joining our Facebook Group: www.facebook.com/groups/survivalpreppingbasics. There, you can let us know how you are doing, ask any questions you may have, or offer advice about your expertise. We wish you nothing but continued success and happy learning!

REFERENCES

Chapter 1

AIP Safety. 2021. "The main objectives and goals of first aid training." Aipsafety.com,. Accessed 10 July, 2022. https://aipsafety.com/news/objectives-and-goals-of-first-aid-training/.

Alvarado, Sara. 2020. "What is the average emergency response time?". Medical News Bulletin. Accessed 22 June, 2022. https://medicalnewsbulletin.com/response-time-emergency-medical-services/.

Bowman, Warren D. 2001. "Perspectives on being a wilderness physician: is wilderness medicine more than a special body of knowledge?" *Wilderness & Environmental Medicine* 12 (3): 165-167.

Charlton, Nathan P., Jeffrey L. Pellegrino, Amy Kule, Tammy M. Slater, Jonathan L. Epstein, Gustavo E. Flores, Craig A. Goolsby, Aaron M. Orkin, Eunice M. Singletary, and Janel M. Swain. 2019. "2019 American Heart Association and American Red Cross Focused Update for First Aid: Presyncope: An Update to the American Heart Association and American Red Cross Guidelines for First Aid." *Circulation* 140 (24): e931-e938. https://doi.org/doi:10.1161/CIR.0000000000000730.

Conger, Cristen. n.d. "How does your brain impact your survival chances in the wilderness?". Science.howstuffworks.com. Accessed 10 July, 2022. https://science.howstuffworks.com/life/inside-the-mind/human-brain-survival-psychology.htm.

Sholl, J. M., and E. P. Curcio, 3rd. 2004. "An introduction to wilderness medicine." *Emerg Med Clin North Am* 22 (2): 265-79, vii. https://doi.org/10.1016/j.emc.2004.01.001.

Chapter 2

Abebe, Mulugeta. 2009. "Emerging trends in disaster management and the Ethiopian experience: genesis, reform and transformation." *Journal of Business and Administrative Studies* 1 (2): 60-89. https://doi.org/10.4314/jbas.v1i2.57352/.

Cunha, John P. . 2021. "What Should Be in an Emergency Survival Kit?". eMedicineHealth. Accessed 10 July, 2022. https://www.emedicinehealth.com/what_should_be_in_an_emergency_survival_kit/article_em.htm.

Fuerst, Ron. n.d. "First Aid Kits - 33 Essentials to Prepare for Emergencies." eMedicineHealth. Accessed 10 July, 2022. https://www.emedicinehealth.com/first_aid_kits/article_em.htm.

Myers, Courtney. 2017. "The Importance of Mental Preparedness Prior to a Major Natural Disaster." The American Red Cross. Accessed 10 July, 2022. https://preparecenter.org/story/the-importance-of-mental-preparedness-prior-to-a-major-natural-disaster/.

Porcelli, A. J., and M. R. Delgado. 2017. "Stress and Decision Making: Effects on Valuation, Learning, and Risk-taking." *Curr Opin Behav Sci* 14: 33-39. https://doi.org/10.1016/j.cobeha.2016.11.015.

Sena, Lelisa, and Kifle Woldemichael. 2006. "Disaster prevention and preparedness." *Ethopia Public Heal Train Initiat* 1: 1-80. https://www.cartercenter.org/resources/pdfs/health/ephti/library/lecture_notes/health_science_students/lln_disaster_prev_final.pdf.

Wemm, S. E., and E. Wulfert. 2017. "Effects of Acute Stress on Decision Making." *Appl Psychophysiol Biofeedback* 42 (1): 1-12. https://doi.org/10.1007/s10484-016-9347-8.

References

Chapter 3

Bennett, Wayne 2019. "First Aid Training: HAINES Recovery Position." Disastersurvivalskills.com,. Accessed 27 July, 2022. https://disastersurvivalskills.com/blogs/preparedness/first-aid-training-haines-recovery-position.

Centers for Disease Control and Prevention. n.d. "Warning Signs and Symptoms of Heat-Related Illness." cdc.gov. Accessed 11 July, 2022. https://www.cdc.gov/disasters/extremeheat/warning.html.

Chen, Joyce. 2017. "Performing CPR to the Bee Gees, Beyonce and Missy Elliot could save lives." Today.com. Accessed 10 July, 2022. https://www.today.com/health/here-s-ultimate-list-songs-help-you-perform-cpr-t112073.

Clinical Quality & Patient Safety Unit. 2019. "Clinical Practice Procedures: Trauma/Manual in-line stabilisation." Accessed 10 July, 2022. https://www.ambulance.qld.gov.au/docs/clinical/cpp/CPP_Manual%20inline%20stabilisation.pdf.

Coyne, Thomas. 2021. *Survival Medicine: The Essential Handbook for Emergency Preparedness and First Aid*. Rockridge Press.

Dictionary.com. n.d. "cardio." Accessed 11 July, 2022. https://www.dictionary.com/browse/cardio-#:%7E:text=Cardio%2D%20is%20a%20combining%20form,many%20medical%20and%20scientific%20terms.

Dowshen, Steven 2018. "Your Heart & Circulatory System." Kidshealth.org,. Accessed 11 July, 2022. https://kidshealth.org/en/kids/heart.html.

EMTResource.com. 2014. "DCAP BTLS." Accessed 11 July, 2022. http://www.emtresource.com/resources/acronyms/dcap-btls/.

Forgey, W. . 2020. *The Prepper's Medical Handbook*. Lyons Press.

Hatraining.com. n.d. "Unconscious Casuality." Accessed 27 July, 2022. https://www.hatraining.com/first-aid/unconscious-casualty/.

Healthline. 2019. "How to Take Your Pulse (Plus Target Heart Rates to Aim For)." Accessed 11 July, 2022. https://www.healthline.com/health/how-to-check-heart-rate.

Healthline.com. 2021. "How to Help Someone Having a Panic Attack." Accessed 13 June, 2022. https://www.healthline.com/health/how-to-help-someone-having-a-panic-attack.

Healthwise Staff. 2020. "Rest, Ice, Compression, and Elevation (RICE)." Uofmhealth.org. Accessed 11 July, 2022. https://www.uofmhealth.org/health-library/tw4354spec#:%7E:text=As%20soon%20as%20possible%20after,Ice%2C%20Compression%2C%20and%20Elevation.

Healthywa.wa.gov.au. n.d. "First aid for spinal injury." Accessed 10 July, 2022. https://www.healthywa.wa.gov.au/Articles/F_I/First-aid-for-spinal-injury.

Lee, K. 2012. "Cardiopulmonary resuscitation: new concept." *Tuberc Respir Dis (Seoul)* 72 (5): 401-8. https://www.e-trd.org/journal/view.php?doi=10.4046/trd.2012.72.5.401.

Mayo Clinic Staff. 2022. "Anemia." Mayo Clinic. Accessed 11 July, 2022. https://www.mayoclinic.org/diseases-conditions/anemia/symptoms-causes/syc-20351360.

Mayo Clinic Staff. n.d. "Sudden cardiac arrest." Mayo Clinic. Accessed 10 July, 2022. https://www.mayoclinic.org/diseases-conditions/sudden-cardiac-arrest/symptoms-causes/syc-20350634?p=1.

Merriam-webster.com. n.d. "pulmonary." Accessed 11 July, 2022. https://www.merriam-webster.com/dictionary/pulmonary#:%7E:text=Definition%20of%20pulmonary,pulmonary%20artery%20a%20pulmonary%20embolism.

Nhsinform.scot. 2022. "First aid." Accessed 10 July, 2022. https://www.nhsinform.scot/tests-and-treatments/emergencies/first-aid#the-recovery-position.

Stöppler, Melissa Conrad 2021. "Medical Definition of Choking (object in airway)." MedicineNet. Accessed 10 July, 2022. https://www.medicinenet.com/choking_object_in_airway/definition.htm.

Thim, T., N. H. Krarup, E. L. Grove, C. V. Rohde, and B. Løfgren. 2012. "Initial assessment and treatment with the Airway, Breathing, Circulation, Disability, Exposure (ABCDE) approach." *Int J Gen Med* 5: 117-21. https://doi.org/10.2147/ijgm.S28478.

Chapter 4

Arnold, R. W. 1999. "The human heart rate response profiles to five vagal maneuvers." *Yale J Biol Med* 72 (4): 237-44.

Bennett, Wayne 2019. "First Aid Training: HAINES Recovery Position." Disastersurvivalskills.com,. Accessed 27 July, 2022. https://disastersurvivalskills.com/blogs/preparedness/first-aid-training-haines-recovery-position.

Betterhealth.vic.gov.au. 2014. "Head injuries and concussion." Accessed 8 June, 2022. https://www.betterhealth.vic.gov.au/health/conditionsandtreatments/head-injuries-and-concussion.

Centers for Disease Control and Prevention. n.d. "Symptoms of Mild TBI and Concussion." Accessed 11 July, 2022. https://www.cdc.gov/traumaticbraininjury/concussion/symptoms.html.

Cheung, N. H., and L. M. Napolitano. 2014. "Tracheostomy: epidemiology, indications, timing, technique, and outcomes." *Respir Care* 59 (6): 895-915; discussion 916-9. https://doi.org/10.4187/respcare.02971.

Fokkens, W. J., V. J. Lund, C. Hopkins, P. W. Hellings, R. Kern, S. Reitsma, S. Toppila-Salmi, M. Bernal-Sprekelsen, and J. Mullol. 2020. "Executive summary of EPOS 2020 including integrated care pathways." *Rhinology* 58 (2): 82-111. https://doi.org/10.4193/Rhin20.601.

Forgey, W. . 2020. *The Prepper's Medical Handbook*. Lyons Press.

Goldman, Rena 2018. "11 Effective Earache Remedies." Healthline.com.

Accessed 12 July, 2022. https://www.healthline.com/health/11-effective-earache-remedies#neck-exercises.

Heimark, Cory 2010. "Wilderness Medicine and Survival Tips." Imminent Threat Solutions. Accessed 19 June, 2022. https://www.itstactical.com/medcom/medical/wilderness-medicine-and-survival-tips/

Kucik, C. J., and T. Clenney. 2005. "Management of epistaxis." *Am Fam Physician* 71 (2): 305-11.

Le Sage, N., R. Verreault, and L. Rochette. 2001. "Efficacy of eye patching for traumatic corneal abrasions: a controlled clinical trial." *Ann Emerg Med* 38 (2): 129-34. https://doi.org/10.1067/mem.2001.115443.

Lockett, Eleesha 2019. "9 Ways to Naturally Clear Up Your Congestion." Healthline.com. Accessed 12 July, 2022. https://www.healthline.com/health/natural-decongestant#neti-pot.

Rana, Sarika 2018. "9 Effective Home Remedies To Stop Nose Bleeding." Food.ndtv.com. Accessed 13 June, 2022. https://food.ndtv.com/health/effective-home-remedies-to-stop-nose-bleeding-1842722.

Reilly, P. L., D. I. Graham, J. H. Adams, and B. Jennett. 1975. "Patients with head injury who talk and die." *Lancet* 2 (7931): 375-7. https://doi.org/10.1016/s0140-6736(75)92893-7.

Roland, James. 2019. "How Are Broken Ribs Treated?". Healthline. Accessed 12 July, 2022. https://www.healthline.com/health/treatment-for-broken-ribs#rest.

Schofield, Kirsten 2017. "7 Natural Remedies for Your Upset Stomach." Healthline. Accessed 12 July, 2022. https://www.healthline.com/health/digestive-health/natural-upset-stomach-remedies

Chapter 5

Alton, Amy 2015. *The Ultimate Survival Medicine Guide: Emergency Preparedness for ANY Disaster*. Skyhorse.

References

American Red Cross. n.d. "Adult First Aid/CPR/AEd Ready Reference." Accessed 28 July, 2022. https://www.redcross.org/content/dam/redcross/atg/PDFs/Take_a_Class/Adult_Ready_Reference_Card.pdf.

American Red Cross. 2022. "Child & Baby CPR." Accessed 28 July, 2022. https://www.redcross.org/take-a-class/cpr/performing-cpr/child-baby-cpr.

Barrell, Amanda 2020. "CPR steps: A visual guide." Medical News Today. Accessed 12 June, 2022. https://www.medicalnewstoday.com/articles/324712#cpr-step-by-step.

Forgey, W. . 2020. *The Prepper's Medical Handbook*. Lyons Press.

Furst, John. 2017. "The Jaw Thrust Technique – a step by step guide." Accessed 12 July, 2022. https://www.firstaidforfree.com/the-jaw-thrust-technique-a-step-by-step-guide/.

Habrat, Dorothy 2019. "How To Insert a Nasopharyngeal Airway." Merckmanuals.com. Accessed 12 July, 2022. https://www.merckmanuals.com/professional/critical-care-medicine/how-to-do-basic-airwayprocedures/how-to-treat-the-choking-conscious-infant.

Health Jade Team. 2019. "Head tilt chin lift." Accessed 12 July, 2022. https://healthjade.net/head-tilt-chin-lift/.

Martin, Paul 2022. "When to Perform CPR – How to Tell if Someone Needs CPR." Procpr.org. Accessed 12 July, 2022. https://www.procpr.org/blog/training/when-to-perform-cpr.

Mayo Clinic Staff. 2020. "Choking: First aid." Mayoclinic.org. Accessed 12 July, 2022. https://www.mayoclinic.org/first-aid/first-aid-choking/basics/art-20056637.

Nicholson, Billie 2015. "First Aid – Sterilizing Medical Instruments." Sun Oven. Accessed 12 July, 2022. https://www.sunoven.com/first-aid-sterilizing-medical-instruments/.

O'Connor, Terry 2016. "SPLINTING REVIEW." Coloradowm.org.

Accessed 12 July, 2022. https://www.coloradowm.org/blog/splinting-review/.

Redcross.org. n.d. "First Aid/CPR/AED Care During COVID-19." The American Red Cross. Accessed 7 July, 2022. https://www.redcross.org/take-a-class/coronavirus-information/first-aid-cpr-aed-care-during-covid-19.

Royallifesavingwa.com.au. n.d. "First Aid Hygiene – Royal Life Saving WA. (n.d.). Royal Life Saving." Accessed 12 July, 2022. https://royallifesavingwa.com.au/your-safety/first-aid/first-aid-hygiene.

Sears, Brett. 2021. "How to Wear a Shoulder Sling." Verywellhealth.com. Accessed 12 July, 2022. https://www.verywellhealth.com/how-to-properly-wear-a-sling-on-your-arm-2696291.

Sja.org.uk. n.d. "How to make an arm sling." Accessed 12 July, 2022. https://www.sja.org.uk/get-advice/first-aid-advice/how-to/how-to-make-an-arm-sling/.

Stang, Debra 2018. "How to Make a Splint." Healthline. Accessed 12 July, 2022. https://www.healthline.com/health/how-to-make-a-splint.

Tjzawesome. n.d. "How to Make a Splint/ Fix a Broken Bone." Instructables.com. Accessed 12 July, 2022. https://www.instructables.com/How-To-Make-A-Splint-Fix-A-Broken-Bone/.

Chapter 6

Bellingar, B. . n.d. "What Constitutes a Minor Emergency? ." Drbellingar.com. Accessed 12 July, 2022. https://www.drbellingar.com/blog/what-constitutes-a-minor-emergency.

Brouhard, Rod. 2022. "10 Basic First Aid Procedures." Verywell Health. Accessed 12 July, 2022. https://www.verywellhealth.com/basic-first-aid-procedures-1298578.

Cronkleton, Emily 2022. "11 Home and Natural Remedies for Toothache

References

Pain." Health line. Accessed 19 June, 2022. https://www.healthline. com/health/dental-and-oral-health/home-remedies-for-toothache

Fanous, Summer 2019. "13 Home Remedies for Mosquito Bites." Accessed 19 June, 2022. https://www.healthline.com/health/outdoor-health/home-remedies-for-mosquito-bites

Forgey, W. . 2020. *The Prepper's Medical Handbook*. Lyons Press.

Furst, John. 2020. "First Aid for a Migraine Headache." First aid. Accessed 12 July, 2022. https://www.firstaidforfree.com/first-aid-for-a-migraine-headache/.

Hepler, Linda 2017. "First Aid for Bites and Stings." Healthline. Accessed 12 July, 2022. https://www.healthline.com/health/first-aid/bites-stings.

Kidshealth.org. 2021. "Bug Bites and Stings." Accessed 19 June, 2022. https://kidshealth.org/en/parents/insect-bite.html.

M, Dr. Sruthi. 2021. "How Do You Break a Fever Naturally?". Medicine Net. Accessed 19 June, 2022. https://www.medicinenet.com/how_do_you_break_a_fever_naturally/article.htm.

Mayo Clinic. 2022. "Fever: First aid." Accessed 12 July, 2022. https://www.mayoclinic.org/first-aid/first-aid-fever/basics/art-20056685?reDate=01032022#:%7E:text=The%20average%20body%20temperature%20is,temperatures%20than%20younger%20people%20have.

Mayo Clinic Staff. 2022. "Fever." Mayoclinic.org. Accessed 12 July, 2022. https://www.mayoclinic.org/diseases-conditions/fever/symptoms-causes/syc-20352759.

MedicineNet. 2022. "How to Cure Mouth Ulcers Fast Naturally." Accessed 12 July, 2022. https://www.medicinenet.com/how_to_cure_mouth_ulcers_fast_naturally/article.htm.

Medlineplus.gov. n.d. "Dizziness." Accessed 12 July, 2022. https://medlineplus.gov/ency/article/003093.htm.

Nhs.uk. 2019. "Treatment." Accessed 12 July, 2022. https://www.nhs.uk/conditions/insect-bites-and-stings/treatment/.

Prescott Dentistry. 2022. "Natural Home Remedies for Jaw Pain Relief." Accessed 19 June, 2022. https://prescottdentistry.com/home-remedies-for-jaw-pain-relief/.

Share.upmc.com. 2021. "The Dangers of a High Fever." Accessed 12 July, 2022. https://share.upmc.com/2016/10/fever-treatment-guidelines/.

Sja.org.uk. n.d. "Fainting." Accessed 12 July, 2022. https://www.sja.org.uk/get-advice/first-aid-advice/unresponsive-casualty/fainting/.

Tagazier, O. 2021. "How to Naturally Get Rid of Mouth Ulcers." Shelton Causbas. Accessed 12 July, 2022. https://sheltoncausbas.blogspot.com/2021/11/how-to-naturally-get-rid-of-mouth-ulcers.html.

Watson, Kathryn. 2019. "10 Home Remedies for Vertigo." Healthline. Accessed 19 June, 2022. https://www.healthline.com/health/home-remedies-for-vertigo.

Chapter 7

Brouhard, Rod. 2022a. "10 Basic First Aid Procedures." Verywell Health. Accessed 12 June, 2022. https://www.verywellhealth.com/basic-first-aid-procedures-1298578.

Brouhard. 2022b. "10 Types of Second-Degree Burns." Verywellhealth.com. Accessed 19 June, 2022. https://www.verywellhealth.com/examples-of-second-degree-burns-1298346#:~:text=Friction%20burns%20don't%20involve,changes%20can%20usually%20prevent%20infection.

Centers for Disease Control and Prevention. 2022. "Lightning Safety Tips." Accessed 19 June, 2022. https://www.cdc.gov/disasters/lightning/safetytips.html.

Electronics-notes.com. n.d. "What is Electric Current: the basics." Accessed

References

19 June, 2022. https://www.electronics-notes.com/articles/basic_concepts/current/what-is-electrical-current.php.

Forgey, W. . 2020. *The Prepper's Medical Handbook*. Lyons Press.

Heimark, Cory 2010. "Wilderness Medicine and Survival Tips." Itstactical.com. Accessed 19 June, 2022. https://www.itstactical.com/medcom/medical/wilderness-medicine-and-survival-tips/

Intermountain Healthcare. n.d. "Chemical Wounds." Accessed 12 July, 2022. https://intermountainhealthcare.org/services/wound-care/wound-care/conditions/chemical-wounds/.

Mayo Clinic Staff. 2022a. "Chemical burns: First aid." Mayo Clinic. Accessed 19 June, 2022. https://www.mayoclinic.org/first-aid/first-aid-chemical-burns/basics/art-20056667?reDate=01032022.

Mayo Clinic Staff. 2022b. "Poisoning: First aid." Mayo Clinic. Accessed 12 July, 2022. https://www.mayoclinic.org/first-aid/first-aid-poisoning/basics/art-20056657.

Mayoclinic.org. 2022. "Burns: First aid." Accessed 27 July, 2022. https://www.mayoclinic.org/first-aid/first-aid-burns/basics/art-20056649.

Medlineplus.gov. n.d. "Shock." MedlinePlus. Accessed 19 June, 2022. https://medlineplus.gov/ency/article/000039.htm.

NHS. 2021. "Concussion." Accessed 19 June, 2022. https://www.nhsinform.scot/illnesses-and-conditions/injuries/head-and-neck-injuries/concussion.

NHS Inform. 2021. "Concussion." Accessed 11 July, 2022. https://www.nhsinform.scot/illnesses-and-conditions/injuries/head-and-neck-injuries/concussion#treating-concussion.

Thebetterindia.com. 2018. "Be Prepared: 10 Common Medical Emergencies & How to Deal With Them." Accessed 19 June, 2022. https://www.thebetterindia.com/155315/first-aid-medical-emergencies-news/.

Unicef.org. 2021. "Caring for someone with COVID-19 at home." Accessed 19 June, 2022. https://www.unicef.org/rosa/stories/caring-someone-covid-19-home.

Webmd.com. 2021. "What Are the Types and Degrees of Burns?". Accessed 19 June, 2022. https://www.webmd.com/first-aid/types-degrees-burns.

Chapter 8

Centers for Disease Control and Prevention. 2022a. "Seizure First Aid." Accessed 19 June, 2022. https://www.cdc.gov/epilepsy/about/first-aid.htm.

Centers for Disease Control and Prevention. 2022b. "Stroke Signs and Symptoms." Accessed 19 June, 2022. https://www.cdc.gov/stroke/signs_symptoms.htm.

Clinicaltrials.gov. 2021. "The Effect of Umbilical Cord Clamping Distance." Accessed 19 June, 2022. https://clinicaltrials.gov/ct2/show/NCT04862403.

The Editors of Encyclopaedia Britannica. n.d.-a. "antibody." Britannica.com. Accessed 19 June, 2022. https://www.britannica.com/science/antibody.

The Editors of Encyclopaedia Britannica. n.d.-b. "antigen." Britannica.com. Accessed 19 June, 2022. https://www.britannica.com/science/antigen.

Health.harvard.edu. 2005. "Emergencies and First Aid - Childbirth." Accessed 19 June, 2022. https://www.health.harvard.edu/staying-healthy/emergencies-and-first-aid-childbirth.

Maximizedchiro.com. n.d. "Options for Pushing During Labor & Delivery." Accessed 19 June, 2022. https://www.maximizedchiro.com/options-for-pushing-during-labor-delivery/.

Mayo Clinic Staff. 2022a. "Anaphylaxis: First aid." Mayo Clinic. Accessed

References

19 June, 2022. https://www.mayoclinic.org/first-aid/first-aid-anaphylaxis/basics/art-20056608.

Mayo Clinic Staff. 2022b. "Heart attack." Mayoclinic.org. Accessed 19 June, 2022. https://www.mayoclinic.org/first-aid/first-aid-heart-attack/basics/art-20056679?reDate=04032022.

McDermott, Annette 2019. "First Aid for Stroke." Healthline. Accessed 19 June, 2022. https://www.healthline.com/health/stroke/stroke-first-aid#for-caregivers.

MedlinePlus. n.d. "Seizures." Accessed 19 June, 2022. https://medlineplus.gov/ency/article/003200.htm.

Mongan, Marie. 2018. "HypnoBirthing: the Mongan method." In *A Natural Approach to Safer, Easier, More Comfortable Birthing*. Audible Studios.

National Women's Health. n.d. "Stages of labour Zoom information sessions." Accessed 19 June, 2022. https://nationalwomenshealth.adhb.govt.nz/womens-health-information/maternity/labourandbirth/stages-of-labour/.

Pietrangelo, Ann 2020. "Allergic Reaction First Aid: What to Do." Healthline. Accessed 19 June, 2022. https://www.healthline.com/health/allergies/allergic-reaction-treatment.

Chapter 9

Ansorge, Rick 2020. "Heat Exhaustion." WebMD LLC. Accessed 19 June, 2022. https://www.webmd.com/fitness-exercise/heat-exhaustion#1.

Cunha, John P. 2021. "Sunburn (Sun Poisoning)." eMedicineHealth. https://www.emedicinehealth.com/sunburn/article_em.htm.

Kidshealth.org. 2018. "First Aid: Frostbite." Accessed 19 June, 2022. https://kidshealth.org/en/parents/frostbite-sheet.html.

Marks, Hedy 2021. "Gangrene." WebMD. Accessed 19 June, 2022.

https://www.webmd.com/skin-problems-and-treatments/guide/gangrene-causes-symptoms-treatments.

Mayo Clinic Staff. 2022a. "Heatstroke: First aid." Mayo Clinic. Accessed 19 June, 2022. https://www.mayoclinic.org/first-aid/first-aid-heatstroke/basics/art-20056655?reDate=06032022.

Mayo Clinic Staff. 2022b. "Sunburn." Accessed 19 June, 2022. https://www.mayoclinic.org/first-aid/first-aid-sunburn/basics/art-20056643?reDate=06032022.

Sja.org.uk. 2021. "Dehydration." Accessed 11 July, 2022. https://www.sja.org.uk/get-advice/first-aid-advice/effects-of-heat-and-cold/dehydration/.

Skinsight.com. n.d.-a. "Frostbite, First Aid." Accessed 19 June, 2022. https://www.skinsight.com/skin-conditions/first-aid/first-aid-frostbite.

Skinsight.com. n.d.-b. "Heat Exhaustion, First Aid." Accessed 19 June, 2022. https://www.skinsight.com/skin-conditions/first-aid/first-aid-heat-exhaustion.

WebMD. 2021. "Drowning Treatment." Accessed 19 June, 2022. https://www.webmd.com/first-aid/drowning-treatment.

Wildmedcenter.com. 2016. "Hypothermia." Accessed 19 June, 2022. https://www.wildmedcenter.com/blog/hypothermia.

Chapter 10

ActionHub Reporters. 2011. "How To Make a Tree-Pit Snow Shelter." Actionhub.com. Accessed 19 June, 2022. https://www.actionhub.com/how-to/2011/06/28/how-to-make-a-tree-pit-snow-shelter/.

Animatedknots.com. n.d.-a. "Clove Hitch – Rope End." Accessed 11 July, 2022. https://www.animatedknots.com/clove-hitch-knot-rope-end#ScrollPoint.

References

Animatedknots.com. n.d.-b. "Lashing Knot – Diagonal." Accessed 19 June, 2022. https://www.animatedknots.com/diagonal-lashing-knot.

Animatedknots.com. n.d.-c. "Lashing Knot – Shear." Accessed 19 June, 2022. https://www.animatedknots.com/shear-lashing-knot.

Animatedknots.com. n.d.-d. "Lashing Knot – Square." Accessed 19 June, 2022. https://www.animatedknots.com/square-lashing-knot.

Animatedknots.com. n.d.-e. "Timber Hitch." Accessed 19 June, 2022. https://www.animatedknots.com/timber-hitch-knot#ScrollPoint.

Bhaddock. 2012. "How to build a Three-Pole Tarp Tepee Shelter." Bildernessarena.com. Accessed 19 June, 2022. https://www.wildernessarena.com/food-water-shelter/shelter-natural-shelter/three-pole-tarp-tepee-shelter.

Finchamp, Deanna 2019. "A-Frame Shelter With Paracord." Sgtknots.com. Accessed 19 June, 2022. https://sgtknots.com/blogs/news/a-frame-shelter-with-paracord.

Fouche, Maz. 2022. "8 Basic Survival Knots You Should Know." Skyaboveus.com. Accessed 19 June, 2022. https://skyaboveus.com/wilderness-survival/8-Essential-Knots-You-Should-Know-Survival-Skills.

Gearpatrol.com. 2014. "The Wilderness Guide to Natural Shelters." Accessed 19 June, 2022. https://www.gearpatrol.com/outdoors/a111521/wilderness-guide-to-natural-shelters/.

Hutchison, Patrick 2016. "How to Tie Lashings." Artofmanliness.com. Accessed 11 July, 2022. https://www.artofmanliness.com/skills/manly-know-how/how-to-tie-lashings/.

Krebs, Jessie. 2017. "Best survival knots every prepper should learn." Theprepared.com. Accessed 11 July, 2022. https://theprepared.com/survival-skills/guides/best-3-survival-knots-every-prepper-should-learn/.

Lung.org. n.d. "Ventilation: How Buildings Breathe." Accessed 19 June, 2022. https://www.lung.org/clean-air/at-home/ventilation-buildings-breathe.

McCann, John D. . 2011. "Snow Trench Shelter." Survivalresources.com. Accessed 19 June, 2022. https://www.survivalresources.com/snow-trench-shelter.html.

Michael. 2021. "Survival In The Wilderness." Ussartf.org. Accessed 19 June, 2022. https://ussartf.org/survival_wilderness.htm.

Outdooradventureguide.co.uk. 2017. "Survival bushcraft: How to build a lean-to shelter and go tent-free camping." Accessed 19 June, 2022. http://www.outdooradventureguide.co.uk/how-to-build-a-lean-to-shelter-and-go-tent-free-camping/.

Raveendran, Rugma 2022. "How Does An Igloo Keep You Warm?". Scienceabc.com. Accessed 19 June, 2022. https://www.scienceabc.com/nature/how-does-an-igloo-keep-you-warm.html.

Ready.gov. 2022. "Tornadoes." Accessed 19 June, 2022. https://www.ready.gov/tornadoes?gclid=Cj0KCQiA2sqOBhCGARIsAPuPK0jWCQQisuSpfLRjC5Jw63FUUEzW-pxXPW2IFEj92QXNXouSo8tTOi-IaAnhREALw_wcB.

Reddit.com. 2018. "How to make an igloo." Accessed 19 June, 2022. https://www.reddit.com/r/coolguides/comments/9goqpv/how_to_make_an_igloo/.

Scoutlife.org. n.d. "How to Build a Snow Cave." Accessed 11 July, 2022. https://scoutlife.org/outdoors/150860/how-to-build-a-snow-cave/.

Smith, Lindsay N. 2017. "How to Survive a Natural Disaster." Accessed 19 June, 2022. https://www.nationalgeographic.com/adventure/article/how-to-survive-natural-disaster-storm-hurricane-expert-tips.

The Survival Journal. n.d. "How To Build a Dugout Shelter." Accessed 19 June, 2022. https://thesurvivaljournal.com/dugout-shelter/.

Telsonsurvival.com. n.d. "How to Build Shelter in the Wilderness (11 Types of Shelter Explained)." Accessed 19 June, 2022. https://telsonsurvival.com/prepping/emergency-preparedness/wilderness-survival-shelters.

Thefreedictionary.com. n.d. "lash together." Accessed 19 June, 2022.

References

https://www.thefreedictionary.com/lash+together#:%7E:text=Verb,their%20victim%20to%20the%20chair%22.

Uwc.211ct.org. 2022. "How to Prepare and Safely Weather a Hurricane/Tornado." Accessed 19 June, 2022. https://uwc.211ct.org/how-to-prepare-and-safely-weather-a-hurricanetornado/.

VertDude. n.d. "How to Build an Igloo Out of Snow." Instructables.com. Accessed 19 June, 2022. https://www.instructables.com/How-to-build-an-Igloo-out-of-snow/.

Chapter 11

Adamant, Ashley 2018. "How to make Dandelion root coffee." Practicalselfreliance.com. Accessed 19 June, 2022. https://practicalselfreliance.com/dandelion-coffee/

Adamant, Ashley. 2021. "How to make Dandelion Oil (& 7 ways to use it)." Practicalselfreliance.com. Accessed 27 July, 2022. https://practicalselfreliance.com/dandelion-oil/

Andrew. n.d. "Does Boiling Water Purify It?". Knowpreparesurvive.com. Accessed 11 July, 2022. https://knowpreparesurvive.com/survival/does-boiling-water-purify/#How-Can-You-Purify-Water-Without-Boiling-it.

Brusie, Chaunie 2019. "What Are the Most Effective Natural Antibiotics?". Healthline. Accessed 19 June, 2022. https://www.healthline.com/health/natural-antibiotics

Cronkleton, Emily. 2019. "Ginger for Arthritis: Should I Give It a Try?". Healthline. Accessed 19 June, 2022. https://www.healthline.com/health/ginger-for-arthritis.

Haigh, J. . 2019. "Wild First Aid Introduction. IGO Adventures." Accessed 19 June, 2022. https://blog.igoadventures.com/2019/09/19/wild-first-aid-introduction/.

Hamilton, Andy 2009. "How to forage a first aid kit." Theecologist.org. Accessed 20 June, 2022. https://theecologist.org/2009/oct/06/how-forage-first-aid-kit.

Health.ny.gov. 2018. "Boil Water Response - Information for the Public Health Professional." Accessed 19 June, 2022. https://www.health.ny.gov/environmental/water/drinking/boilwater/response_information_public_health_professional.htm#:%7E:text=The%20standard%20recommendation%20for%20boiling,kill%20or%20inactivate%20waterborne%20pathogens.

Indian Country Today. 2018. "Natural Pain Relief: 9 Alternatives to Ibuprofen, Acetaminophen or Aspirin." Nativeknot.com. Accessed 19 June, 2022. https://www.nativeknot.com/news/Native-American-News/Natural-Pain-Relief-9-Alternatives-to-Ibuprofen-Acetaminophen-or.html.

Kps. n.d. "Staying Clean in the Wild: The Natural Way." Knowpreparesurvive.com. Accessed 19 June, 2022. https://knowpreparesurvive.com/survival/wilderness-hygiene-hacks/#Organic-Flu-Medicine.

Michael. 2021. "Survival In The Wilderness." Ussartf.org. Accessed 19 June, 2022. https://ussartf.org/survival_wilderness.htm.

Modernhippiehabits.com. n.d. "Foraged Medicinal Herbs First Aid Kit." Accessed 19 June, 2022. https://modernhippiehabits.com/index.php/2016/07/14/medicinal-herbs-first-aid-kit/.

Newsnetwork.mayoclinic.org. 2020. "Turmeric's anti-inflammatory properties may relieve arthritis pain." Accessed 19 June, 2022. https://newsnetwork.mayoclinic.org/discussion/mayo-clinic-q-and-a-turmerics-anti-inflammatory-properties-may-relieve-arthritis-pain/.

Plantaddicts.com. n.d. "Are Elderberry Poisonous?". Accessed 19 June, 2022. https://plantaddicts.com/are-elderberry-poisonous/.

Sarnacki, Aislinn 2019. "10 Medicinal plants for your natural first aid kit." Hellohomestead.com. Accessed 19 June, 2022. https://hellohomestead.com/10-medicinal-plants-for-your-natural-first-aid-kit/.

INDEX

A

Abdominal pain 61, 63, 94, 95, 96, 156
Alexander Selkirk 14
Allergic reaction 187
Anaphylaxis 189
  Symptoms 64, 87, 150, 152, 188, 189, 190, 219
  Treatment
    EpiPen 152, 191
Allergies
  Anaphylactic shock 166
    epinephrine 167
Anemia 50
Automated external defibrillator 186

B

Bleeding 169
  Hypovolemic shock 170
  Treatment
    direct pressure 170
    Neo-Synephrine 170
Bones
  Dislocation(s) 97, 98
  fracture(s) 5, 9, 60, 76, 89, 92, 93, 97, 100, 111, 112, 161
    fem 97
    femur 98
    traction 97, 98

Breathing
  ABCDE 36, 39, 58, 59, 64, 65, 71, 74, 80, 119, 127, 180
  Abnormal 45
  Airway 28, 33, 36, 39, 40, 41, 43, 44, 45, 46, 47, 77, 79, 80, 118, 119, 120, 121, 124, 125, 126, 127, 129, 131, 132, 153, 169, 219
    infant 139
  Airway blockage 135
  Airway, opening 28
  Anaphylaxis 117, 152, 189, 190
  Antibody 187
  Antigen 187
  Artificial respiration 28
  Breathing exercises 93, 196
  Choking 45, 89, 117, 126, 131, 132, 134, 139, 143, 191, 192, 195
    infant 139
  Burns 158, 175
    Severity
      First-degree 175
      Third-degree 175
    Treatment 175
    Types 175

C

Carotid massage 91
Chemical Wounds 171

305

causes 171
Seek help first 173
Treatment 172
Chest trauma 61, 92
Child birth 192
  3 stages of labor 202
  Breathing exercises 202
  Breathing Techniques 194
  Cutting the Cord 205
  deliver positions 196, 197, 198, 199, 200
  hypnobirthing 193
  Rules 193
  To Push or Not to Push 196
Concussions 173
  First aid 173
  Symptoms 173
COVID-19 ix, 3, 70, 180, 181

D

Defibrillator 117, 186
Dental pain 84, 148
Diagnostic Techniques 59
Dizziness 146, 161, 189
  Treatment
    Epley maneuver 161
Drowning 218
  Treatment 219

E

Ear/hearing Problems 84
Electrocution 118, 176
  Know the power source 177
  Severity 177
  Treatment 177
Environmental exposure 49
Eye/Vision issues 81

F

Fainting 146, 153
FEMA
  Disaster supply kit 18
Fevers 154
  prevention 155
  symptoms 155
  treatment
    medications 157
    treatment 157
First aid 7, 8, 12
  Airway blockage 135
  Broken nose 89
  CPR 14, 26, 27, 28, 41, 44, 46, 58, 59, 74, 81, 106, 116, 117, 118, 122, 123, 126, 127, 130, 131, 135, 138, 140, 142, 152, 153, 167, 169, 170, 177, 180, 186, 187, 191, 204, 214, 217, 218, 219
  Direct pressure 71
  faint 153
  Heimlich 106, 131, 132, 134, 138, 143
  Insect bite
    stinger removal 152
  level of consciousness 47
  Migraine 147
  Nosebleed 89
  Pulse 51, 66, 67
    brachial 54
    carotid 52
    children 57
    infant 57
    popliteal 56
    radial 53
  Respiratory Rate 68
  RICE 71, 103

Index

Scene Safety 31
Sling 20, 101, 111, 112, 115, 161, 294
Splints 62, 63, 77, 103, 106
First-aid kit 18, 20
Foreign bodies 87
   nasal speculum 88
Fractures
   broken ribs 92
Frostbite 217
   First aid 218
   Symptoms 217

G

Gall bladder stones 94
Gastritis 94, 95

H

Headache 61, 63, 147, 155, 156
Head injury 76
Heart attack 185
   First aid 185
   Symptoms 185
   Treatment
      AED 186
Heat exhaustion 71, 154, 211
   First aid 211
   Symptoms 211
Heatstroke 212
   First aid 213
   Symptoms 213
Home remedies
   Dental pain 148
      Hydrogen peroxide mouthwash 148
      Salt water mouthwash 148
   Essential oils 87
   joint pain
      magnesium 150
   mouth sores 159
   Neti pot 86, 87
   Saline irrigation 86
Hypothermia 50, 72, 214
   First aid 215
   Symptoms 214
   Treatment 216

I

Insect Bites and Stings 150

J

Joint Pain 102

K

Knots 223
   Bowline Knot 229
   Bowyer's Knot 234
   Clove Hitch 233
   Clove hitch knot 233, 235, 237, 239, 246
   Double Overhand Knot 224
   Double Sheet Bend 232
   Figure 8 Follow 228
   Figure 8 Knot 226
   Figure 8 Loop 227
   Figure 8 on a Bight 227
   Flemish Knot 226
   Flemish Loop 227
   Halter Hitch 231
   Overhand Knot 223
   Reef Knot 225
   Round Turn and Two Half Hitches 230
   Square Knot 225
   Thumb Knot 223
   Timber Hitch 234

L

Lashings 235
  Diagonal 237
  Shear 239
  Square 235
Lightning 178
  Treatment 179
Lightning Safety 178, 179

M

Medication
  acetaminophen 87, 147, 157, 174, 181, 269
  Alternatives
    Antibiotics 268
    Painkillers 269
  Antacid 95
  ibuprofen 21, 172, 174, 210, 218, 268, 269
  paracetamol 87, 147, 157, 172, 174, 181, 210, 218, 268
Medicinal plants
  Arnica 273
  Bearded lichen 278
  Burdock 273
  Calendula 274
  Cattails 274
  Common plantain 275
  Common yarrow 275
  Dandelion 276
  Goldenseal 277
  Jewelweed 277
  Lavender 278
  Old man's beard 278
  Selfheal 279
  White Willow 279
Migraine 147
  triggers 147

Mouth Sore 158
  treatment 159
Mouth ulcers
  causes 158

N

Natural Disasters 263
  Hurricane 264
  Stay alert 263
  Tornado 264
Natural remedies 85, 86, 95
  Apple cider vinegar 96
  Chamomile 95
  Elderberries 281
  Ginger 85, 95, 269, 280
  Heating pad 96
  jointpain 149
  Peppermint 95, 149
Nature's Bandages 281
  Birch Polypore 282
  Seaweed 281
Neck injury 78
Nose Problems 86
  Nasal Congestion 86

P

Poisoning 118, 168
  Symptoms 168
Pregnancy 192
Psychological strength 10, 16

Q

Quarantine 180

R

respiratory 68

Index

S

Salmonella 116
SAMPLE 44, 64, 65, 74
Seizures 76, 191
  first-aid 191
Septic shock 166
Shelter 221, 222
  Where to build 240
Shelters
  Types 241
    A-Frame 246
    Debris Hut 259
    Dugout 261
    Igloo 255, 258
    Lean-To 242
    Ridgepole 259, 260
    Snow Cave 252, 256, 257
    Snow Trench 249
    Tarp 244
    Tarp tepee 244
    Tree Pit Snow Shelter 258
Shock 166
  causes 166
  Phases 167
    compensated phase 167
    Progressive/decompensated phase 167
Signaling 262
Sprains 160
Sterilization 114
Stress 16, 158, 288
Stress management 150
Stroke 186
  Treatment 187
Sunburn 209
  Treatment 210
Sun poisoning
  Symptoms 209
Survival Medicine Vs. First Aid 7

Survival treatment
  Reduction 99, 100

T

Tachycardia 90
Testicle pain 100
Testicular torsion 101
Twisted testicle 102
Typhoid fever 70, 114, 116

U

Ulcers 94, 158

V

Vagus nerve 91
vertigo. *See* Dizziness
Vital signs 64, 65. *See* First-Aid:- Pulse
  blood pressure 10, 52, 65, 68, 69, 70, 74, 90, 92, 155, 167, 189
  respiratory rate 44, 45, 46, 58, 68, 74, 117, 275
  Temperature 66, 70

W

Water 18, 178, 271
  Purification 272

Made in the USA
Columbia, SC
08 July 2024